Starting Points for a Healthy Habitat

Editing by AveryHartley Marketing and Communications.

Typography, format and cover design by Mel Typesetting.

Original "healthy house" and "sick house" line art by Debra Bond.

Coloring of houses on cover and other original line art
by Joanie Barbier Studios.

Illustrations on pages 56-60 and 117 by Ed Nies.
All photos by the author
Printing by Millennia Press.

Disclaimers: This book contains no medical advise and should not be substituted in any way for the appropriate medical care of a healthcare professional. The opinions and conclusions expressed are solely by the author and in no way or manner are meant to be specific advice for the treatment of any condition, medical or otherwise. In fact, the whole point of the book is that when authoritative sources are conflicting or incomplete, each individual must take personal responsibility for their own health and well-being. Personal stories are composites drawn from the author's experience of his clients and do not apply to or identify any one specific individual.

ISBN 0-9671525-0-X

Library of Congress Catalog Card Number: 99-63090

Published by GMC Media, Denver, CO

Table of Contents

ACKNOWLEDGMENTS

The seed of the idea for this book was planted by Helen Poff, more than six years ago. She was introducing me as the speaker for a local support group, the Rocky Mountain Environmental Health Association. She concluded her remarks by quietly asking me when I was going to write my book. I was dumbfounded. The idea had never occurred to me. In fact, it was preposterous! Yet, a tiny, but very clear, voice in my head said that someday I *would* write a book.

Just prior to that incident, I met Robert Ivker, D.O., author of the book *Sinus Survival.* We developed a friendship, discussed goals and plans and agreed to create a business. Although the joint business never came to fruition, our individual ones did — each heavily influenced by the other. In addition, his generosity in including me in the third edition of *Sinus Survival* has created exposure for my endeavors that no marketing firm can duplicate. The readers of his book who have called me for help and guidance have consistently demonstrated to me that stable, intelligent people across the country are hungry for information about stopping indoor exposures — and are equally frustrated at finding appropriate, effective guidance that has their best interest in mind.

Throughout all this, Nicholas G. Nonas, M.D., was quietly becoming a mentor. At first, he was the primary person responsible for my regaining my health, both through his medical care and his superb referrals. Then he encouraged me to use what I had learned in my own experience to help others. As we began to collaborate, we explored new ideas and tools for identifying, measuring and mitigating sources of indoor allergens. We both were aware that the particular combination of his medical care and my indoor assessments and mitigation recommendations were becoming more and more effective. His patients were deriving increasing benefit from our collaboration. We wanted to expand our ability to help.

Our efforts to develop and execute a variety of formal and informal studies led to a pilot project based on the allergen measurement methods of Dr. Robert Hamilton of the Johns Hopkins University. We then applied that experience to writing a grant proposal. We didn't get the grant but the process of formulating and clarifying our actions led directly to several key concepts within this book.

Dr. Nonas then asked me to co-present a continuing education course for his medical academy. This led to my writing a simple booklet two years ago for the doctors taking the class, to provide guidance in working with their patients to stop exposures. Their response was very positive, but with a suggestion: Rewrite the book for use by the patient. I didn't need a "tiny inner voice" to tell me what kind of book I would write.

As I look back over the past 15 years, I am humbly amazed at the influence of Dr. Nonas on my methodology and attitudes toward my clients — and on the direction my life has taken. If there is one fundamental and necessary force behind the recovery of my health and discovering the salvation gained from helping others, it is Dr. Nonas.

On a tangential note, Mary Starkey, of the Starkey International Institute of Household Management, asked me to teach a class as part of their curriculum. Her belief in my message, and the patience of her students over the past eight years, has allowed me to present, develop and rehearse my message to a very different audience. I was faced with the task of educating people with little personal experience with life-altering events from indoor exposures to contaminants and about how to identify and address the problem in others. This challenge was critical to the development and writing of this book.

There would be no book, of course, without the thousands of clients over the past dozen years who have so generously opened their houses — literally — to my probing eyes, nose, ears and interminable questions. It is they who have taught *me*. It is they who have trusted my initial trial-and-error approach — and risked the possibility that their complaints would not improve — that led to the development of my guided trial-and-error procedures, and eventually to the systematic approach of open feed-back loops developed in this book. It is also they who have enhanced my experiences by relating additional information about what worked and what didn't. The personal stories in the text are either based on my client's direct statements or are the synthesized composites of many of them. This book is the accumulated experience and wisdom of all my clients.

Special thanks to: Ann Fisk, a friend and colleague from the beginning; Ken and Maggie Dominey, not only for their video (see Chapter 1), but for their example of dealing with adversity in the midst of giving to others; Robert Rinehart, Sr., Ph.D., of Rinehart Laboratories, Inc. for his interest and curiosity in my desire to measure airborne microorganisms in a different way, and his assistance in achieving that; Debra Bond for the original drawings of the "Happy House" and the "Sick House"; Joanie Barbier Studio for all the original line art and final cover illustration; and to Jeanine Corrigan who patiently and laboriously edited my initial draft. All I can say is that I'm glad the manuscript wasn't being formally graded. I would have flunked by any measure, whether "strictly by the curve" or by any modification thereof! Barbara and Art Ratkewicz provided invaluable feedback on many of the ideas contained in the early manuscript. The credit for the development of the map and travel analogies belongs to them.

I would also like to thank Steve Olson, D.D.S., for his encouragement, trust, jokes and amazing mercury-free dentistry practice; David Trevarthen, M.D; and Pam and Frank Benison. In addition, my spiritual teacher is a living example of infinite and unconditional trust, support and love. His example allowed me to alter key victimization behaviors created by my life experiences and which were subsequently amplified by this typically insidious and non-obvious syndrome.

Special appreciation goes to Gay Lasher, Psy.D., David Passikov, Deane Shank, Ph.D., and Nicholas Nonas, M.D., who so graciously and enthusiastically contributed their interviews.

Thanks to Pam Avery of AveryHartley Marketing and Communications, my editor, for her professionalism and her devotion to honoring my voice in the midst of re-writes, re-organizations and deletions. Despite the fact that she was very concerned that I would take her critiques personally, she clearly stated them. I took them personally only once — and she was still correct. Her guidance about single spaces after periods, commas, "em dashes" and clarity of writing directly increased the effectiveness of my message by quantum leaps.

Ed Nies of Mel's Typesetting had to deal with a different way of educating me. Namely, how to prepare a book so that it *can* be typeset.

My son, Jon, had a special role in this saga. At one time I felt that I had lost all in my life but him. And right then in the midst of his own exposure difficulties, he wanted (and needed) his independence. Imagine that! Granting him that gift turned out to be an extraordinary gift to not only him, but to myself as well. My mother made this book possible in innumerable ways, including two key criticisms — wouldn't you know *that* would come from a mother! — that were turning-points for making my message available to a broader audience.

Special hugs, kisses and love to two remarkable children, Courtney and Spencer. Their gifts of love, trust and play are priceless, fun and life-altering. (They also generously allowed me the use of their computer when mine was unavailable!) And to their magnificent mother, Lin, I offer this: How strangely wonderful the twists and turns of life can sometimes be. Without my susceptibility and the damnable exposures and the subsequent but *slooooow* healing, without the influences that led to my creating a consulting company to help people with similar exposure problems, without your brother and his doctor, and lest we not forget a certain unnamed contractor, we never would have met and become friends. And now our paths have turned again, become convergent and synchronous, transforming from healing and surviving to thriving. And then it evolved again to become, as you envisioned, creative. That was a new starting point which allowed the completion of this book. Thank you, my darlin'.

INTRODUCTION

Do your allergies, asthma, headaches or fatigue worsen whenever you are in your house? Do you generally feel sick and tired indoors? Could it be that your workplace or home is making you feel ill? If you aren't sure, or if you have been unsuccessful at stopping your complaints, then you most likely have encountered a variety of frustrating obstacles for a task that most everyone else claims should be extremely simple.

Perhaps your doctor has suggested a thorough housecleaning. But, in your opinion, the house is in pristine condition and you resent his implication that you are a lousy housekeeper. Or maybe, you would like to clean the house, but even a simple vacuuming gives you a migraine. Perhaps you hired a professional company to clean your furnace ducts and your carpets but your symptoms are no better than before. You suspect the culprit may still be in the house or office itself. But when you bring this up, nobody — not public health authorities, medical professionals, insurance companies or even your own families and friends — seems to believe you.

Do you believe in yourself?

You may be among the growing number of people whose health and well-being have been impacted at levels varying from mild "hay fever" type symptoms, to interference with your lifestyle or even the extreme of a life-altering disability. Yet you are rarely able to receive responsible, effective — or even *compassionate*—personal support and medical care. Those who feel they are harmed by indoor exposure to nontoxic, noncancerous — and therefore non-regulated — substances, often have a hard time convincing others of their claims.

Are you convinced of your claims?

Furthermore, as I know from personal experience and that of most of my clients, the greater the impact the more difficult it is to pursue your own recovery — and the more likely that those friends who no longer share your experiences will begin to lose patience and even desert you. And then, those of you who *persist* in *demanding* assistance often are considered hypochondriacs, labeled neurotic or even diagnosed as mentally ill. Most others, especially authorities, do not easily accept the unfamiliar problems of others as being real.

Are you willing to persist on your own?

If so, here are three fundamental obstacles that must be understood and then overcome:

- Inadequate, misleading or conflicting information about what to do, how to do it and what to avoid.
- A personal support system that has gradually become inappropriate or perhaps hostile.
- The unchallenged belief that the necessary public health and safety standards are also *sufficient* for individuals.

These obstacles are what can sometimes make the seemingly simple task of cleaning up your house or office so difficult and even aggravating — to yourself and everyone around you.

THE START

Despite these formidable obstacles, you must start somewhere. So, lacking guidance or professional procedures — and instead of standing idly by hoping for conditions to change — start wherever you happen to be. And then be prepared, when you don't get the results you anticipated, to begin again. We are not talking about a procedure that has been previously clarified and simplified, like inserting tab A into slot B. Rather, we have to first define "tab," "slot," "A," "B" and "insert."

Because there are no previously established procedures that apply to *your* situation, you will have to create them yourself. You do this by experientially creating a feedback system. You make your best estimate as to the problem, decide on an action, observe the results, evaluate and begin again. You continue this loop, refining your actions each time, until you are satisfied and choose to stop. If you are motivated to improve your health and well-being, and if you are actively involved in the process, any mistakes you make can be eventually discovered and corrected.

The ***validation*** of your efforts will be the elimination of, or at least the reduction of, your complaints. Your experience then has the potential to be used in later group studies by experts. But first, you may want to take care of yourself.

This book is designed to teach by example. It will "loop" you through the various components of the subject matter several times. But each presentation is from a slightly different point of view. The purpose is to teach you a feedback process of inquiry, decision-making, action and evaluation so that you can successfully fine-tune *your* personal formula to:

- Identify both the physical and psychological factors of your complaints.
- Find and generate the information you need to make good decisions about effective actions.
- Evaluate the results.
- Start again to further improve your situation.
- Educate your personal support system and public authorities so they can better assist you.

THE DIFFICULTY

The primary difficulty in identifying and solving indoor air quality problems is not the lack of more sophisticated technology, nor is it the lack of a miracle drug. The vast majority of failures result from incomplete — or even misleading — information about the details of what *actually* happens indoors rather than what we *believe* happens. In other words, it's a case of previously accepted knowledge and "mind-set" over the truth.

These beliefs are based on a combination of cultural and authoritative factors. The cultural factors are primarily "common-sense" beliefs that are passed on by word-of-mouth — such as what constitutes a clean house — and from industry sources that must meet only minimal legal requirements governing marketing claims and product labels.

The authoritative factors are primarily the demands of regulatory compliance agencies with their subsequent legal liability issues, public health standards for catastrophic events affecting large segments of society, and our reliance on "experts" to define our experience and direct our behavior.

These beliefs then appear to be validated by our trust in our advanced technology. We forget that the events happen to *people* in the real world, not in a test tube under controlled laboratory conditions. We also tend to assume that because technology is so powerful — much of which is beyond our comprehension and thus appears to be almost magical — that it can detect and measure *everything*.

Conversely, if an individual's experience cannot be proven by using advanced technology, non-sufferers and the authorities respond as if nothing actually occurred — which can only mean that our experience, in their expert opinion, was either mistaken or was illusory. And if we persist in our claims? Then, obviously, we are hysterical, or worse.

Any sufficiently advanced technology is indistinguishable from magic.
Arthur C. Clark

Any sufficiently personal experience is indistinguishable from illusion.
Carl Grimes

Any sufficiently persistent claim of a personal experience is indistinguishable from hysteria.
Carl Grimes

However, even if these political, social, cultural and regulatory mind sets are resolved, this problem of what-to-do and what-to-avoid will still persist. It will persist until manufacturers of products and providers of services provide ***full and truthful disclosure*** of ***all*** information concerning their products and services that you and your healthcare provider deem necessary to successfully protect your health and well-being.

This book presents details about what information is necessary and how to evaluate it for *your* needs. Appendix B is an extensive listing of books and Internet resources that presents even more information and resources. Unfortunately, if the providers of products and services that generate exposures choose to limit their information to the minimum legal requirements, then you can only do the best you can — mistakes and all — with what you have. Or, as computer aficionados are fond of saying: "Garbage in, garbage out."

The difficulty behind what should be a relatively simple problem of stopping indoor exposures to contaminants and irritants is not any *one* of the above issues — or even a combination of them. The difficulty lies in the fact that all are related to each other in a system of *closed* feedback loops. A closed feedback loop sees only what is in the loop. The information that the loop generates only reinforces and validates what is already in the loop. It rejects all else as irrelevant anomalies, misguided personal bias, artifacts of unavoidable human involvement or worker's compensation fraud.

This closed system of information is the primary reason these relatively simple problems have not already been solved.

THE HOPE

The hope lies in opening the feedback loop, going outside the assumed boundaries of conventional wisdom and common sense, at multiple points of entry. The hope lies in an *open-ended* inquiry into personal experiences of indoor exposures to nontoxic, noncancerous and non-regulated substances. It lies in looking and seeing what *actually* happens rather than quickly *judging and categorizing* an event as either "correct" or "crazy." But most of all, the hope lies with people — those individuals who perceive and validate their personal experiences of events, and those who persist in attempts to fix any resulting harm.

Fortunately, some people *do* observe and respect the unique and unpredictable experiences of others as being real and significant, even if they don't fully understand. Many even have had success in improving the health and well-being of those suffering from these seemingly untreatable ailments. Healthcare professionals with this attitude of acceptance and compassion use the authority of their status to help others, rather than to dispute their claims. For them, the burden of proof is on themselves to discover what is happening to their patient, rather than it being on the patient to prove they are suffering.

What these remarkable healthcare professionals have in common is that they don't limit their knowledge and actions to the technical procedures within their specialty. Instead, they take an integrated, whole-person, approach to diagnosis and treatment tailored to each person's unique and *total* situation.

They have the attitude that just because medicine has no standard diagnosis for a patient's complaint doesn't mean that *nothing* is happening nor does it automatically mean that the patient is uncontrollably hallucinating or willfully perpetrating some sort of fraud. They understand that there is more to life and healing than stubborn adherence to a standardized process and to the rules of regulatory compliance.

Likewise, there are people suffering from individual exposures who have stopped relying on those authorities who insist that their experience is "all in their heads." They have rejected others' tendency to "blame the victim" and sought out healthcare professionals who really do care about health — *their* health. They are willing to open their mind to new or different information and processes, but only when they have a need. In the meantime they can ignore or deny irrelevant information like the rest of the public. They have also found other individuals like themselves and have explored their commonalities and differences. Their scouting reports from the "front lines," so to speak, reveal a territory rich in future research possibilities.

But most importantly, at this initial stage of discovery, they have taken a stand on the legitimacy of individual experiences. They insist that just because your experience doesn't strictly match the official guidelines doesn't mean it is wrong. Just because your experience is different from the majority of society doesn't mean you should be stigmatized. Just because the public can't see or feel your experience doesn't mean that you have to remain helpless. It just means that *you* have to take responsible action to protect yourself and to educate others, rather than rely on someone to do it for you. You also have a responsibility to help open the closed-loop of belief.

This position also implies — actually it *mandates* — that the product manufacturers, service providers and public authorities fulfill their equal responsibility to supply you with *complete* and *accurate* information so you *can* take care of yourself. Anything less further erodes an already shaky support system and *creates* more of the victimization they claim they are trying to prevent in the first place — if only you would stop creating your own problems for yourself.

THE PLEA

It is also a plea to those of you who successfully end your complaints to actively and vigorously educate non-sufferers and public authorities alike. Don't become silent and disappear, leaving the impression that nothing really happened to you, and that the others who still are suffering are just neurotic. Tell your story! Don't hide from the fear of being associated with "those sickies." Instead, dispel the fear that others have — through your persistence, education, information and example.

SUGGESTIONS FOR READING THIS BOOK

The subject matter of this book is based on my experience and opinions. It does not require comprehending esoteric scientific concepts or advanced technology, nor does it rely on magic or New-Age nuances. In fact, conventional methods of executing plans and understanding common terms such as "clean," "safe" and "health" will suffice in the vast majority of situations for most people. Our concern is not with when these conventional methods and plans *do* work, it is when they *don't* work. That is when the advice and information you seek often becomes confusing and conflicting, increasing your frustration.

Then, when conventional methods and plans *consistently* fail, you will begin to experience an additional form of frustration. This psychological and social form of frustration lies with the repeated failures of your best efforts to stop the nagging, or even debilitating, complaints. It increases with the agonizing pursuit of information and advice that always seems to be designed for the best interest of someone else. It lies with the loss of credibility you have with family and friends. It lies in the mounting financial cost of experimenting with products, services and medical care, and the resulting lost time and opportunity.

As you grow more determined and become actively involved in attempts to stop your complaints, your frustration will most likely magnify as you desperately try to make decisions based on an increasingly lengthy list of products, services and "recipes" for action that can be mechanically utilized (see Appendix B). *How do you make choices about what to do and what to avoid? Do you have to do **everything** that is suggested? What can you leave out? What happens if you make a mistake? **How do you create and then maintain a constructive plan in the midst of conflicting beliefs, fears and authoritative claims? How do you manage the often compulsive extremes of pretentious denial and hysterical hypervigilance?***

This book provides some answers to these questions based on my experiences and opinions about why the problem of indoor contaminants exists in the first place, and why solving those problems can sometimes be so aggravating. The answers are presented within the framework of a *customized process* rather than as a universal regimen. The process will generate a personalized plan designed to assist, guide and support you as you generate your own information to best use the available products, services and "recipes" to end suspected complaints caused by exposure to indoor contaminants. The plan requires ***accuracy, trust*** and ***timeliness*** of the following:

- The ***meaning*** of the information.

  A variety of "common-sense" beliefs are illustrated and demythologized. Their errant assumptions are then replaced with new meanings and understandings of familiar processes. If you do not understand or trust the information, it is valueless at best and confusingly harmful at worst.

- The ***choosing*** of what to do and what to avoid.

  Your Personal Impact Rating (PIR), is a self-evaluation method for establishing which exposures sources affect you the most and how they compare to others. Your PIR will be your main guide for establishing priorities when making choices about what to do and what to avoid at various levels of diligence.

- The ***execution*** of the plan.

  Products and services are explained, along with illustrations of how to work with service providers, family, and friends to ensure that the results of your plan are what you actually intend.

- The ***evaluation*** of the results.

  Descriptions about how to evaluate whether or not the plan was successful — what to do if the complaints are worse or if the results prove insufficient.

- The ***choosing*** of the next series of actions.

  How to use the information generated by the previous successes and failures to determine your next starting point.

The personal experience you gain from repeating this action-loop until the desired results are achieved, is the key to successfully reaching your goal of a complaint-free indoor environment. Therefore, you can use this book however you choose. Start where you think best and stop when you desire.

Ignore what you believe is irrelevant and focus on what you are convinced is critical. In other words, "if it ain't broke, don't fix it."

It is also important to keep an open mind and to question "common-sense" beliefs and procedures. But if you aren't getting the results you desire — if it *is* broke, for *you* — then you might want to "fix" it. Your success will most likely require new information and new meanings for your experiences — or else you would have already stopped your complaints.

The more simple situations will usually require intervention for only the physical identification and removal of sources. If that is your situation, then you may find the social and psychological factors irrelevant. However, if your initial process fails, then the management of the social and psychological factors will most likely become necessary. If further attempts fail, "loop" back through the process until you either generate a satisfactory solution or conclude that it is more cost effective to move to a new house.

Finally, use this book in conjunction with appropriate medical care, not instead of it. This book is not meant to replace testing, scientific fact, public health and safety laws or established medical procedures. All of these **are** important and should be complied with *first*. Rather, this book is meant to provide guidance for extending and enhancing that information by:

- Filling the gaps when objective or scientific data is unavailable or insufficient.
- Personalizing data so that *you* find it meaningful and usable.
- Suggesting ways to take *positive action* to successfully reduce, if not eliminate entirely, your complaints.
- Initiating a discussion about what to expect and how others have dealt with non-obvious exposure problems whose impact alters your life expectations.

Again, when scientific and medical authorities claim that they can no longer help you, that there is no scientific or physical basis for your complaints, that is your clue to act on your own. You don't have to wait for science to catch up to you. You can still take responsible, effective action to significantly reduce — if not eliminate — your complaints about indoor exposures to sick buildings, common allergens, respiratory irritants, unpleasant odors, and asthma inducing triggers.

Chapter One

My Starting Points

My Starting Points

I was an ideal patient. I followed my doctor's instructions to the letter, even taking all my antibiotics, rather than quitting as soon as I started feeling better. And I believed in their expertise and authority. If anyone's opinion differed even slightly from the established, I denounced it as "quackery."

Unfortunately, loyal belief and following instructions weren't enough to keep me well. About 14 years ago, I began noticing that several common health complaints had become more numerous and uncomfortable. I could no longer ignore them. Aspirin, exercise and antibiotics no longer relieved frequent headaches, muscle fatigue and sore throats. Visits to the doctor became more frequent.

As a stubbornly independent person, I had always "toughed it out" and assumed that I could continue to do so. However, my symptoms grew steadily worse. Often my mind felt "foggy" until midday, preventing me from working comfortably or with joy. I was too tired to participate in the simple activities I used to enjoy such as jogging, bicycling through the park, going to a movie, or even eating a favorite dinner of steak with all the trimmings. At the office, my work suffered. As my short-term memory began to fail me, I made mistakes and missed appointments.

Meanwhile, I began to notice that I didn't feel well after I was in certain buildings, including my own office and home. "Don't be ridiculous!" I told myself. I had never heard of buildings making anyone sick, and neither had anyone else I asked. As the "strep-like" sore throats grew worse, my doctor gave me more antibiotics. When those stopped working, he prescribed a stronger antibiotic. Eventually, I was on antibiotics virtually all the time. Within 48 hours of finishing a prescription, the flu-like symptoms and sore throats would return with a vengeance, only to disappear again within a few hours of starting another dose of antibiotics.

At this point, I feared for my life. I was convinced that the "germs" in my body were becoming resistant to the stronger antibiotics. And I became outraged when my doctor referred me to an *allergist*! I thought, "How would treating hay fever, which I don't have, kill off all those dangerous germs?" I refused to go. About that time, more than one person suggested that I was a hypochondriac and that my real problem was an immature, neurotic need for attention.

FIRST SUCCESSFUL HELP

Fortunately, my doctor didn't agree with them. He convinced me that the allergist he referred me to was also experienced with lesser known side effects from antibiotics. Reluctantly, I went — a move that changed my life.

The allergist understood what I was experiencing and knew what to do about it. What I found out about the role of antibiotics and environmental exposure started me on a remarkable journey.

I learned how antibiotics can sometimes result in an overgrowth of yeast organisms by unbalancing the natural microbial flora of the intestinal tract. And I learned about new diets and anti-fungal drugs in addition to antibacterial medications. (By the way, in the past 14 years I have needed conventional antibiotics only twice — despite recent double-blind, controlled studies to the contrary.)

I also discovered the "total load" concept and how the yeast toxins could overwhelm my body's natural responses, resulting in an avalanche of secondary problems. It was during this period that I was introduced to environmental exposures and their effect on my body.

At first, I thought that the yeast problem was the sole cause of my complaints and that I had no exposure problems. However, as the yeast was brought under control, I began noticing strong reactions to auto exhaust, perfume, cleaning agents, solvents, mold, and even some of my favorite foods.

I then learned about exposure "overload" and "masking" — how some exposures, including yeast from inside the body, are like a smoke-filled room. If a room is full of smoke all the time, it becomes the "normal" condition and you don't connect your suffering with the constant irritant. Minor changes from other sources go unnoticed in the "normal haze." Not until you clear the air do you notice that there had been smoke in the room and that it was actually coming from many sources. As I became adept at avoiding the sources that bothered me, I felt even better.

Did the environmental exposures lead to the yeast problems or was it the other way around? Was it my genetics? Or did the exposures and/or yeast damage my body to begin with? I'm still not sure which came first or what caused what, nor does it matter. What I do know is that thanks to a superb referral for medical care and by learning what exposures to avoid, my symptoms gradually diminished and I began to regain my health. Even more important, I understood more about what was happening to me in a way that made sense, felt right and I could successfully accomplish.

That foundation of reliable knowledge and the resulting comprehension of my treatment was critical. Without it, I might easily have been overwhelmed by the worry and stress of being nearly disabled and unable to work for two years, getting a divorce, and losing my business and house. With it, I was able to successfully maintain my treatment despite those overwhelming events and debilitating environmental exposures.

The other significant event in my healing was a short video by Ken and Maggie Dominey. Maggie was near death with this malady, but she has since slowly recovered to lead a near-normal life. Both she and Ken were members of the Rocky Mountain Environmental Health Association, a local support group. Their video documented her extraordinary struggle to eliminate the sources of exposure that caused her severe suffering. Her courage and determination left me with four key impressions:

1. *This illness is real and should be treated very seriously.*
2. *Something can be done about it.*
3. *I was not alone.*
4. *Although I had been victimized, I didn't have to stay a victim.*

BACK TO WORK

I took the latter realization to heart. Encouraged by a local doctor, I founded an indoor environmental health consulting company, Healthy Habitats®.

During my first two to three years of consulting, I did a lot of research and experimentation. There were so few similarities among clients' houses that I saw each situation as totally unique. About the only thing clients had in common was that they were suffering and that there had to be something I could do to help.

My mission was to discover what that "something" was. But first I had to learn more. However, I found few books on the subject, the notable exception being *An Alternative Approach to Allergies*, by Theron Randolph, M.D. I talked to laboratory scientists, industrial hygienists and health department officials. They were sympathetic, but primarily concerned with regulatory compliance. They knew of few regulations that applied to residential problems. I talked to microbiologists, ecologists, environmentalists, chemists and allergists. Again, none had applied his or her knowledge to the personal indoor environment. Eventually, I found some people who were doing "home audits." However, they typically had a vested interest in selling specific products or services rather than meeting the needs of their clients. Again, a dead end.

During this search, I came across Dr. Robert Rinehart, of Rinehart Laboratories, Inc. He took an active interest in indoor environmental health problems and helped me develop a simple, inexpensive test for airborne mold and other microorganisms. This eventually led to a comparative database of indoor microbial levels that has become a valuable guide for evaluating my clients' indoor environments.

Later, I discovered the work of Dr. Henry Vyner in his book *Invisible Trauma: The Psychosocial Effects of the Invisible Environmental Contaminants*. Dr. Vyner studied the major historical incidences of toxic exposure, such as Three Mile Island, Love Canal, and the bombing of Hiroshima, and analyzed the psychological and sociological effects upon the victims. He described the victims of those events as perceiving that they had been *personally* harmed but unable to obtain *public* confirmation. His findings, especially his model of denial-vigilance-hypervigilance, became my inspiration and working model as I further developed my methods.

With this broader base of knowledge and some elementary tools, simple patterns started to emerge. I began to generate ideas about what sort of conditions could influence the problems my clients were reporting. I then had them make fundamental changes in those conditions while I closely observed the results.

This *blind* trial-and-error technique eventually developed into *guided* trial-and-error. It slowly became more predictive and I was able to more quickly determine what actions had the best chance of success for a specific person with their unique combination of reactivity to their specific exposures. Then I was able to apply my information and experience in a preventive way for pre-purchase inspections or for those not yet suffering any noticeable ill effects. I have even applied my experience to enhancing houses.

Finally, I noticed that as the complaints of the susceptible occupants decreased, the non-complaining occupants also noticed an improvement in their experience. Although they previously had not had a problem, they reported that they now were more comfortable, which led me to speculate that what is marginally safe for the hypersensitive may also be *very* safe for others. And perhaps that could be a powerful adjunct to the standards that safeguard the public in general — especially since the necessary modifications are often relatively minor.

MY NEXT STARTING POINT

The realization and clarification of several key concepts proved necessary before I could develop a successful, predictable system of understanding and a method of attack. I learned by experimenting with literally hundreds of ideas and techniques, and by actively listening to what my clients claimed had and had not contributed to their improvement.

The fundamental starting point for successfully working with my clients was that their experience was real and that their previous efforts demonstrated their motivation and their personal involvement for finding a truthful resolution of their complaints.

Starting with any assumptions other than those would quickly lead to the same dead-end conclusions reached by all the authorities and advisors I had previously consulted. However, with these assumptions as a general starting point, I was able to "see" what was happening and develop a plan of action that even the worst of the so-called hypochondriacs could understand, execute and evaluate. And because they did, that seemed to me to be the strongest proof possible that their complaints were legitimate, that the causes and influences originated somewhere other than in their overactive imaginations or were being generated by some deep-seated psychological malfunction.

Starting with these fundamental assumptions also revealed to me — and to my clients — that the outcome of their efforts were greatly influenced by the understanding, timeliness and accuracy of the information they used as the basis for their decisions and actions. Change their understanding of the information and their choices changed — as did the subsequent results.

These assumptions were fine for when *I* worked with my clients. But attempts to train others in my methods inevitably failed. Something was still missing. Only after several more years and a myriad of situations was I able to clarify other important principles essential to communicating to others what I was intuitively doing.

EXPANDED STARTING POINTS

This book is a presentation of my opinions based on what I have learned from my own experiences and from those of my clients. They are based on the following key points. I hope you find them helpful in developing your own starting points for your own journey.

Use new information and processes in conjunction with *appropriate* medical care, not in place of it. Not all exposure symptoms are caused by environmental factors. You may also have an underlying medical condition whose symptoms are overlapped or mimicked by indoor environmental exposures.

When you reach the end of *public* health care methods and cleanup techniques, yet still feel sick, you have not reached the end of the road. You are actually just at the start of your *individual* journey. What is needed is not a power struggle with those authorities to convince them of the legitimacy of your complaints. What *you* need most of all is a process that guides *you* as *you* travel *your* own path by creating *your* own road map for this critical and often lonely pilgrimage.

Public health and safety laws are designed to benefit the majority of society at a reasonable cost to society. They are necessary and must be complied with *first*. Public health and safety officials do a remarkable job of protecting us from infectious diseases, cancer-causing substances and physical harm. It's a big job and they are to be commended for it.

However, public law has a number of inherent problems:

- In its most grotesque form, it bases decisions on the question of how many lives can be acceptably lost under current conditions until the cost of those lost lives is greater than preventing the losses to begin with.
- It does not address most nonlethal conditions such as prevalence of asthma, allergies, chronic headaches, fatigue, or intolerance of odors and fragrances.
- Assurances, requirements and even edicts of safety based on *public* policy are only a starting point for the *individual*. Again, public law must consider all of society — not the needs of a specific individual. Information *you* can trust is the primary requirement for successfully identifying and solving an individual exposure problem. This is the premise for all the recommendations and techniques presented in this book.

Beware, at least at first, of specific product and service recommendations. Such recommendations, especially in more complex situations, nearly always end up being counterproductive and increase the already inherent confusion. The journey to health typically requires much more than a "magic bullet" product or service. If you have struggled with this issue, chances are your buying a "quick fix" hasn't stopped your suffering from indoor exposures any more than buying a better hammer will help you build a better house.

What is needed, and what this book supplies, are new meanings to common beliefs which then lead to specific, individualized guidelines for using simple tools effectively and economically to (in order of effectiveness):

- **Remove** sources.
  They aren't present anymore.
- **Isolate** sources.
  They are still present but aren't in contact with you.
- **Dilute** sources with ventilation.
  They are still present but at lower levels.
- **Reduce** sources with filtration.
  They are still present but at lower levels.

Note: Some of the details, especially in Chapters 8 and 10, may seem overwhelming at first. If you feel that level of information is unnecessary, skip what you don't need. If you decide later that you do need the more detailed information, you can always revisit those chapters. And you will do so from a different point of view and with a deeper level of understanding.

Generate your *own* information in a way that you understand, trust, and can use when you need it. That is the key to solving *individual* environmental exposure problems. You must perceive the need and have the motivation to personally create your own individual system of discovery, action and evaluation of results. No one can do it for you. You have to take risks with your personal choices and then live with the results — sometimes in direct defiance of someone else's "common sense" or legal requirements, and despite the lack of social acceptance. Anything less leads to a reliance upon the authority of experts to determine meaning and to direct your choices about how you live your life — which increases the opportunity for being victimized.

Begin with the *total set* of environmental parameters. Because the effects of *specific* exposures are not well identified, understood or delineated, you must start with what you have: an individual who has multiple reactions in certain locations or situations. Your task is to determine what can be done to fundamentally change the "total set of parameters" so that your reactions stop completely, or at least become tolerable. Because you are motivated to truly stop your complaints, you will keep only what is of value and discard the rest — even if you have cherished that belief for years. As you continue this process, you will inevitably clarify, fine-tune and self-correct your knowledge and results.

The most appropriate word that describes the *total set* of environmental parameters is *"habitat."* *A human indoor habitat is comprised of the environment, the occupants and all interactions among all the components of the habitat.* This includes the physical exposure events, the individual susceptibility of the various occupants, the interactions between the occupants and the environment, and all interactions — physical, social and psychological — among the occupants themselves. This is a very dynamic situation with multiple variables that constantly change. It has little in common with laboratory settings and controlled experiments. In fact, the mere act of observing an indoor human habitat alters the dynamics that are the target of any investigation.

An evaluation system is needed to prioritize these multitudes of influences that can impact not only your health but especially your understanding of what is happening to you. The lower the rating, the less the impact and the simpler (usually) the solution. The higher the impact rating, the more complex and difficult will be the path to a solution. (See Chapter 7.) An additional benefit to an evaluation system is to provide a comparative context within which you can better understand the changed meanings of your life's situation.

The standard Cause-and-Effect model is not the *only* model for understanding events. Cause-and-effect is the standard model for discovering, among other things, any public health or safety issue. Something happens (a cause) and a result follows (the effect). For example, jumping from an airplane without a parachute will kill you. There is an objectively observable connection between the cause and the effect that is consistent over time, place and groups of people. In fact, it is so consistent that if a cause happens, the effect can be assumed. If the connection isn't observable in some individuals in the test group, that data is discarded because it doesn't fit the model. It's considered an anomaly. And unless its existence is statistically strong enough to change or negate the standardized model, nothing changes. The *status quo* remains.

However, when anomalies occur, especially with a specific individual, that does not mean that ***nothing** happened.* Something ***did*** happen. And it is very real to the person experiencing it. For example, actually surviving a jump from an airplane without a parachute does not mean that nothing happened. It does not mean that the one who jumped hallucinated the event. That individual certainly did jump and certainly did survive. The event did happen and it was extraordinarily real to that person. However, an anomaly occurred; that specific individual survived. No one may ever understand why or how, but that does not disprove the occurrence of that anomaly. An anomaly just means the accepted model isn't applicable in *all* instances, including that specific instance. Neither does it disprove the accepted model. It is merely an exception, an anomaly.

The Influence-and-Tendencies model may also be helpful for discovery and description when an event occurs but the effects aren't uniformly consistent across time, place and individuals. The original action, in other words, did not *cause* a specific effect, but it certainly did *influence* a *tendency* toward future events. Using the parachute example again, the anticipated cause-and-effect sequence didn't occur as the accepted model predicted. But there is still important information to be gathered and used. Even though the cause of the survival may never be known, diligent investigation could most likely discern various influences and tendencies of events, such as wind direction, a muddy landing spot and the fall being "broken" by tree limbs. Perhaps that information could be used to change survival instructions in other emergencies, despite the fact that this particular event may never be duplicated, even by the original individual. In other words, anomalies have an intrinsic value. They need not be abruptly discarded as if they were some awful creature from the "black lagoon" of society.

Finally, an integrated approach to the problem will increase the potential for success. This means integrating medicine, personal action and responsibility, technology, psychology, friends, family and other loved ones, and some form of spirituality into your search for the solution to your health problems. This cannot be successfully delegated to anyone else — not professionals, authorities, construction foremen, industry advisors, public health officials, parents or spouses. You must, and can, do it yourself. In fact, the mere act of creating a feedback loop of accurate, timely information, followed by deciding, acting and reevaluating is itself a healing activity.

In summary, keep in mind that your current methods have failed. This means it may be necessary to consider additional information and perhaps a different way of understanding "common-sense" conditions and events. Some of your assumptions may be false, particularly those you consider to be "normal." You may even discover that the least suspected exposure has the greatest negative impact on your well-being.

By reading the text and then experiencing actual changes in your home, you should be able to generate the *personal* information *you* need, when *you* need it in order to solve *your* exposure problems. As *you* develop *your* plan using *your* available resources (see Appendix B for additional ones), *you* will learn to understand, and then create, *your* own unique, healthy habitat.

CHAPTER TWO

YOUR STARTING POINTS

Your Starting Points

The purpose of this book is to assist you in stopping unwanted exposures to toxic chemicals, sick buildings, common allergens, respiratory irritants, asthma-inducing triggers, and anything else you don't want to be exposed to.

The goal is to create a complaint-free indoor environment. This will be accomplished by determining:

- What to do.
- What to avoid, so you don't replace one problem with another.
- How to evaluate the results.
- How to determine your next starting point.

But first you need to establish a starting point. If you don't know where you are, how will you ever know how to get to where you want?

In fact, you have to establish your own *personal* starting point, discovering where *you* are. All other guidance applies only to large groups of people, such as the general public. There is little established guidance available for individual complaints and no regulations that apply to individual situations.

The only sources of information about individual complaints come from other individuals, such as family and friends. But they are different from you. What works for them may not work for you. In fact, what works for them may make you ill. And if your support system is hostile towards your condition, they are the very ones who may label you as being a hypochondriac.

In addition, you will have to establish a *series* of starting points as you meander through the process of discovering what exactly your complaints are, their possible sources, how to stop the exposure from those sources, and how to determine whether your efforts have been partially or totally successful.

After you establish your first several starting points, you may notice that not only are your points changing, but they are heading in a particular direction. It's as if they begin forming a certain pattern, like connecting the dots in a children's book. Connect enough dots and eventually you can identify the picture. Or to put it in terms of this book, begin with what you know at your current location, try something, see where you are then, and adjust your plans accordingly.

Lacking guidance applicable to yourself is like having a page full of dots with no numbers attached. You can connect any dot to any other dot but no total picture forms. That is meandering blindly. What you need is to meander with guidance.

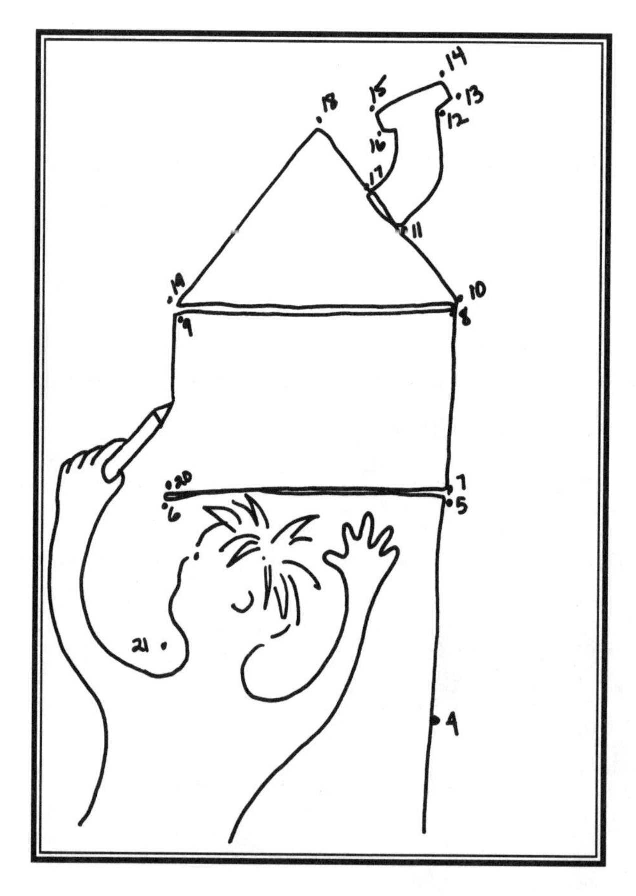

Without guidance and without a plan, where do you start? What's the second step? Where and when do you end? Even with guidance, you may need to draw many pictures before you achieve what you want. And as you experiment, you may change your mind about your final "picture."

This requires establishing a starting point, based on your current knowledge and beliefs, by following these six steps:

Step 1. Locate "where you are." Write down your best description of your current situation and what you want to accomplish by reading this book. Then, specifically, what is the problem or complaint you want to remove?

Step 2. Make a list of all your perceived complaints and symptoms. Be very specific. Do your symptoms vary with the type of exposure? Or do you have just one set of overall symptoms regardless of what you are exposed to? If you aren't sure, that's fine. The purpose of this exercise is to establish a starting point, not solve the whole problem on the first attempt.

Step 3. Rearrange the list of complaints, placing the highest priority first. If you can stop just one complaint, which one would it be? Which one is next?

Step 4. Write down your best estimate as to the most likely cause of the problem. Can you identify a specific source of exposure? Perhaps the key is an event such as a water leak, installing new carpet, or getting a new pet for the kids.

Step 5. What is your best estimate about how to stop that exposure? Fix the water leak, remove the carpet, get rid of the cat, or clean your house again?

Clue: Make note of where you *have* removed possible sources of exposure. Then consider where you *haven't.* The most common locations of *accumulated* exposure sources are:

- Carpets.
- Bedding, upholstered furniture and window coverings.
- Ventilation ducts.
- Walls and ceilings.

Step 6. Decide on how you will know if your actions have been successful. Will you rely on a lab analysis, the assurances of the salespeople who sold you the product, the opinions of your parents or friends, a diagnosis by a medical doctor, or the absence of symptoms?

This evaluation step is critical. It provides the self-correction factor to the process. Your complaint begins with you and does not end until you no longer experience it. The more you have suffered, the more motivation you have to discard beliefs that no longer support your journey. Your desire for the truth will be your guide to improved health and well-being.

If completion of these six general steps achieves your goal of creating a complaint-free indoor environment, congratulations! You are done and you don't need the rest of this book.

However, if you are not satisfied with your initial results, chances are you will want additional information and modifications to common techniques. Keep in mind the six steps above as you progress through your own series of starting points. As you become aware that you need more information and different meanings for "common-sense" assumptions, refer to the appropriate chapters of this book. Chapter 15 presents a comprehensive plan that requires more investigation, new skills and creative problem-solving.

Six Steps to Your First Starting Point

Step 1: Specify the problem or complaint you want stopped.

Step 2: Inventory your symptoms and complaints.

Step 3: Prioritize your symptoms and complaints.

Step 4: Guess the *most likely cause* of the symptoms and complaints.

Step 5: Guess the most likely action to remove the cause.

Step 6: Decide how you will evaluate the results.

Chapter Three

A Medical Starting Point

A Medical Starting Point

A critical component of any starting point is the medical diagnosis and proposed treatment of your complaint. Most physicians have a certain process they follow when faced with complex medical situations. But much depends on how familiar the physician is with treating patients whose health problems defy easy explanation. Following are the observations of one physician who is extremely knowledgeable about environmental and other factors that can influence a person's health and well-being.

Nicholas G. Nonas, M.D. is a family physician who has specialized in environmental medicine for nearly 30 years. He is certified by the American Board of Family Physicians and is a Fellow of the American Academy of Otolaryngic Allergy. Dr. Nonas has extensive experience treating patients who suffer from unconventional conditions such as fibromyalgia, chronic fatigue immune deficiency syndrome, environmental illness and other ailments as yet unnamed.

***Carl Grimes**: Dr. Nonas, how did you become involved with environmental exposure in medicine?*

Dr. Nonas: It was serendipitous. I had a patient whose asthma could be triggered by mold exposure. She also skin-tested positive to mold. But after she had a fire in the basement of her home, her asthma improved. This seemed very curious to me. Why would a fire affect her asthma except to maybe make it worse? I have had other patients with asthma who had house fires and their asthma didn't improve. What was different?

The primary difference was that the fire restoration company had used an ozone treatment as part of their work. A literature check about ozone for *any* use, and especially for exposure to people, was basically negative. I checked with a microbiologist about ozone and learned that he would periodically sterilize his lab with ozone. Ozone was evidently capable of killing mold.

Here was a patient who received more benefit from stopping her exposure to mold than she did from any specific medical treatment.

***CG**: How do you identify patients who will benefit from stopping their exposures from environmental sources, such as the patient you just mentioned?*

Nonas: The patients most likely to benefit from stopping their exposures to allergens are the ones with the most persistent symptoms and the poorest response to conventional medical treatment. These are the ones where I need to look beyond the suspected diagnosis to something else.

***CG**: What treatment methods are most effective?*

Nonas: The treatment methods currently available are to limit the exposure, use prescriptive or natural medication to block the symptoms, or to desensitize the patient so their resistance improves. Of the three, the most successful intervention is avoidance, because if the patient is not exposed, there will be little response that needs treatment.

***CG**: One of the more difficult areas to understand, in my experience from the environmental side, is the variety and complexity of each client's situation. How do you, from the medical side, sort through all this?*

Nonas: Well, it helps to realize that a lot of the differences between individuals are related to genetics. Some are more predisposed than others to be harmed at differing levels to a variety of exposures.

First is the allergic response. It has a precise definition, which includes such language as "IgE-mediated response." It is what most doctors are familiar with and what most allergists focus on.

Next is the inflammatory response, where the body tissues are so irritated and sensitive that they now react strongly to even minor exposures. Next is the toxic condition, where the body is being directly harmed by the exposure.

In addition, there is the priming response, where the irritated membrane becomes primed, or conditioned, to respond to other agents. The body now responds to a variety of agents rather than just specifically to the original sensitizing one.

Another factor is chronicity, or the chronic nature of the exposures. The more frequent or longer term the exposures, the greater the chances that the sequence of sensitization, inflammatory and priming response will intensify.

It is analogous to when your checking account goes into overdraft. You may have just deposited $2,000, only to receive an overdraft notice. The one check that put your account into the red isn't necessarily the key to what happened. The debit trigger can be as small as a $1 if your account is already at the brink. What is most important is how the whole situation evolved, what caused your account to be on the brink of exhaustion so that even a slight debit put it into an overdraft condition.

CG: *It sounds like it could be quite difficult to successfully diagnose and treat some of your patients, especially when their clusters of symptoms don't fit standardized procedures. When the complaints and medical history are complex and highly individualized, how do you sort through the details and keep your focus?*

Nonas: It sometimes is difficult. And it is very easy to just label patients who present a bewildering array of symptoms as neurotic, especially if they are persistent or aggressive at seeking relief from their symptoms. However, complex symptoms to me simply means that I am observing the end of a complex series of events. It's like looking at a giant snowball at the bottom of the mountain. Everything is all jumbled and tangled. That doesn't mean the snowball is neurotic; it only means I'm seeing the events after a lot of history rather than during the early stages of development. The *history* of the snowball is now critical. I use the history to discern how the snowball started, what was the first initial symptom, when it occurred, and under what circumstances.

CG: *How would a patient, or the readers of this book, use this information?*

Nonas: It would help them to go back in time to the beginning of the series of events, rather than try to untangle the resulting "snowball at the bottom of the hill." How did the snowball evolve over time and across events? The answer often lies in that evolution. Look for when you first noticed the response level and then the hypersensitivity level, and how it may have been primed by other events, and determine if it has reached the toxic, or damage, level. Many of the troublesome chronic illnesses I see, such as chronic fatigue, fibromyalgia, and chemical sensitivities, are the snowball at the bottom of a *big* hill. Maybe we can't roll the snowball back to the top and return that individual to his or her original health, but we can usually get it part way back. And that understanding of the evolution of events helps to reduce the emotional and psychological stress associated with trying to cope with the unknown.

***CG**: Are we talking about a traditional medical problem, where there is a specific condition in the body that responds to a specific medicine or surgical procedure?*

Nonas: Sometimes, but usually it is more like having a stone in your shoe. The more you walk, the more sore, or even damaged, your foot becomes. One response is to administer a painkiller so you can keep walking. Another response is to take off the shoe and remove the stone. Many people don't like that approach because if you take a shoe off a damaged foot you may not be able to put it back on immediately. They don't want to wait for the foot to heal. They want a quick fix with a prescription instead.

Many of the conditions I see can be helped with a lifestyle change. I didn't say cured, because the lifestyle didn't cause the problem. But it could be a major factor. For example, with pollen allergies it would be helpful to stay indoors, away from the source of pollen, as much as possible. Keep windows closed, change your clothes upon entering the home, and shower before retiring for the night to remove the pollen from your face and hair. None of those activities will cure the actual allergy, but it will calm the response by reducing the exposure to the pollen.

In our clinic we talk about a pathway to health. The patient, in partnership with the doctors and their staff, work toward discovering where that path is, where it began, and where it could lead them. The focus is then on developing that pathway, looking for the most positive effect first, and then fine-tuning it.

***CG**: How do you respond to someone who says, "Just give me a pill. I don't have time for all this 'New Age mumbo-jumbo.' I want **real** medical treatment."*

Nonas: I'll give them the pill, *if* there is one. Plus, I'll counsel them about the other options. "Real" medical treatment is more than prescribing medications and alleviating symptoms. It is driven by several factors, including the amount of discomfort perceived by patients, their motivation to heal, and their desire to be preventative. As doctors, we have to be prepared to work effectively anywhere along this pathway.

Again, medical treatment at its most effective level is not mechanical like fixing a car engine or a leaky water pipe. It is a partnership between the patient and the medical staff. And most importantly, the treatment has to be "doable" by that particular patient. If patients don't understand the medical directions, then they cannot comply, and the treatment will fail. That breaks the trust between the patient and the doctor, precluding further effective treatment.

Another difficulty arises when multiple issues present themselves. Then the patient, with the guidance of the medical profession, must prioritize the issues — which one to work on first, and which is next. This is where the partnership and its attending issues of open communication and mutual trust is so important. The patient may want to start with issue "A" but the doctor knows that it is caused by condition "B." Therefore, they need to start with "B." If the patient doesn't understand or doesn't trust the doctor, then the treatment plan could fail.

Likewise, if the doctor doesn't understand that his job is to help the patient discover how the snowball developed rather than attempting to untangle it by only treating the symptoms, his guidance will not be sufficient, the complaints won't resolve, and his trust with the patient will suffer.

CG: *How do you identify a patient who may be the victim of a "snowball at the bottom of the mountain"?*

Nonas: First, by applying the technique of the "many syndrome":

Many symptoms.
Many doctors.
Many diagnoses.
Many treatments.
Many failures.

Second, by observing that a change in symptoms occurs with a change in exposure — with the realization that the change, particularly in a complex or advanced case, may not occur until the inflammation stage decreases, which may not occur for as long as 7-10 days after the removal of the irritating stimulus.

***CG**: Back to the "many syndrome," for a moment. In terms of the effect on patients experiencing the "many" failures, what can they do to minimize the emotional and psychological damage that usually results?*

Nonas: I'm not a psychiatrist, and they would have much more to say about this than I can. (See Chapter 17.) But I see the greatest *negative* effects on patients when they blindly accept someone else's belief about their illness — especially when that belief doesn't actually apply to their situation. On the other hand, the *positive* emotional and psychological impact is the greatest when the treatment belief system fits the factual and belief *experience* of the patient. Again, that's why the trust of a doctor-patient partnership is so important; so both sides clearly understand what is happening, what works, and what doesn't.

***CG**: What typically happens when a medical opinion doesn't uncover what the patient is experiencing, especially for illnesses like chronic fatigue syndrome and multiple chemical sensitivities.*

Nonas: The patient will usually be aware that the treatment isn't working. It is helpful for them to guard against continuing a medical relationship if it becomes dysfunctional. And if it does, they should seize control by seeking other opinions and making alternative choices about their care and treatment.

***CG**: One of my challenges is helping my client understand why I can't just test the quality or safety of the indoor air, and then "just fix it." Is there a similar challenge on the medical issues?*

Nonas: There sure is. Patients often ask my staff to test them for their overall health. Or more precisely, they want a blood test for illness. I have to ask them which illness are they concerned about? There are so many that the cost for all of them would be prohibitive. And we may still not have a test available for what ails them. And just because they all return with a negative result, assuming we know what "that" is, doesn't mean they are in perfect health.

Testing is greatly misunderstood. To have a useful test, you first have to understand the mechanism of the event — the clinical diagnosis — before it can be tested. The test can then be useful in confirming the clinical diagnosis. Testing *does not identify* the determining event — what started the snowball rolling down the mountain. Testing *can only confirm* that it came down. Testing can only answer a question that first has to be asked. Testing, in and of itself, says very little without clinical correlation.

Further, we have to understand what a "false negative" result is. In its simplest terms, a false negative is a test result that *incorrectly* says there is nothing wrong. Another way of saying this is that the condition being looked for has not been detected. The danger of a false negative test result is that the underlying mechanism could still be present but we stop looking for it. The damage then continues.

***CG**: How accurate can testing be, under ideal conditions, especially in terms of the false negative result?*

Nonas: The total answer involves complex statistical analysis. But put simply, no test procedure is absolute. The accuracy rate is a balance between the *specificity* of the test and the *sensitivity* of that test.

The more *specific* the test, the fewer conditions it will see; possibly it will detect only one specific condition with the exclusion of all others. In those many negative results some may very well be true. The condition may be present despite the false test result. It just means that this specific test was blind to it. A different test is needed to detect it.

An example of a very specific test is the one for strep throat. The doctor's instructions to the lab is to detect the absence or the presence of the strep bacterium. Therefore, any other organism will be ignored. That information may be important for the proper diagnosis of the patient's total health condition, but the test protocol was too specific to allow it to be visible.

On the other hand, the more *sensitive* the test, the more events it will see, including many "false" ones. These are called "false positive" results. There is a result and that result is positive — but it not may not be true. Using the example above of the strep throat test, the difficulty here is that if the test protocol is too broad, then untold thousands of dollars and hours of time will be spent on identifying the myriad organisms present in any throat culture.

Testing for medical conditions that don't yet have a mature diagnostic history — such things as multiple chemical sensitivity, chronic fatigue, fibromyalgia or even sporadic asthma attacks — typically produces a prevalence of negative results. The difficulty for the medical personnel is the lack of a reliable process to separate the false negatives from the true negatives.

***CG**: Are there any tests for exposure to environmental substances?*

Nonas: I know of one useful test. It was developed by the Johns Hopkins Medical University, Dermatology, Allergy and Clinical Immunology (DACI) Reference Laboratory (see Chapter 10). It asks patients to collect a settled-dust sample using a special collection kit that attaches to their own vacuum cleaners. The sample is sent to the DACI lab for analysis. The results are then sent to the prescribing doctor. The test can detect the allergens from cat dander, dog dander, cockroach, dust mite, mouse urine and mold.

***CG**: How do you use the results in your practice?*

Nonas: If a patient's skin-test is positive to one of the above allergens and the DACI lab results are positive, then I refer him or her to someone who is expert at identifying the actual sources in their home or office, and who can guide the patient toward the successful removal of that allergen. Once the exposure stops, the patient's body will stop being irritated and inflamed. At that point he or she can better respond to medical treatment. That improved state of health usually means that subsequent exposures may not result in an intense allergic reaction like before.

***CG**: Given all the complexity and opportunities for error, how do you work with patients who still can't achieve sufficient relief to experience a comfortable, enjoyable life?*

Nonas: The total answer is beyond medicine alone. An integrated approach that includes other important factors for the comfort and enjoyment of life is necessary. I use three categories of influence: the biochemical sphere of influence, the biomechanical sphere, and the spirituality/psychological sphere. And they usually overlap.

An example would be a patient suffering from headaches. The headaches resulting from biochemical events may be relieved by re-balancing the biochemistry of the body or perhaps blocking part of the pain mechanism. But if the cause is from a whiplash injury in a car accident, that is biomechanical and requires a quite different approach. The third category, that of the spirituality/psychological sphere, has much to do with how that patient responds to the condition of his or her life in total, including the medical treatment.

This is one of the more complex factors in medicine. There is much debate about the mind-body connection. I need to be aware of it, integrate it into my practice, and have good referral sources for my patients. I need experts who understand the psychological *effects* of chronic pain and illness, rather than viewing the psychological dynamics as the *cause* of the physical condition.

***CG**: How does one start the process of "untangling the snowball"?*

Nonas: One must start with the history and then apply what I call "The Allergy Equation" (see Chapter 4). Look for the connections between the susceptibilities and the exposures. Then break that connection by either removing the source of the exposure or improving the susceptibility factor. The difficult cases often require both actions.

CHAPTER FOUR

COMPLAINT EQUATION

Complaint Equation

One specific and necessary starting point is the *Complaint Equation.* It is based on an *Allergy Equation* devised by Nicholas G. Nonas, M.D. It states that if you are experiencing an allergic reaction, then you have to be both allergic to something and exposed to it. If you are allergic, but are not exposed, then you will *not* have a reaction. Nor will you have a reaction if you are exposed but are not allergic.

Sounds simple. Yet it is the fundamental basis for understanding the incredible variability of reactions that occurs with *any* type of toxic, non-toxic, allergic, irritation or sensitivity exposure. It helps remind us that there is more than one way to prevent allergic reactions. You can change your susceptibility or you can change your exposure. Also, changing *both* susceptibility and exposure not only is powerful but often necessary for persons with severe reactions.

"Susceptibility" and "exposure" can each occur at a wide variety and combination of levels. For example, if a person is extremely susceptible to cat dander, then only a small exposure is required to trigger an allergic reaction. That exposure could easily occur just by being near a person who has a cat at home. The trace levels of cat dander clinging to clothing may provoke an allergic attack. Those with moderate susceptibility may not react unless they are in direct contact with a cat. And persons with no susceptibility may never have a reaction, no matter how many cats live in their house.

The ***Complaint Equation*** of this book is an expanded version of Dr. Nonas' *Allergy Equation.* Although still applicable to allergies, the equation now includes respiratory irritants, toxic exposures, asthma triggers, and anything else that involves the combination of "susceptibility" and "exposure."

Susceptibility x Exposure = Subjective Complaint

Not all reactions are allergic reactions, and not all problems are medical in nature. If you are susceptible and exposed, an adverse experience will occur. You perceive a problem and complain. It matters not whether it is an allergic reaction, a toxic exposure, or a broken leg. It is an unwanted experience. You have a complaint. It may be medical and it may not be. It may be an allergic reaction and it may be something else, like an episode of chronic fatigue.

Also, the complaint is not resolved until you perceive that the problem no longer exists. This is true even for obvious medical problems. For example, if your eyes itch and your nose constantly runs, you have a complaint. You seek medical help from an allergist because you believe they are the best resources for that problem. After several visits and a prescription drug, your nose stops running. But your eyes still itch. Everyone agrees your nose has stopped running. But your eyes still itch and only *you* notice it. You still perceive the problem and you will pursue whatever means possible to stop the complaint, even if the doctor insists and a lab test confirms that the diagnosed medical problem can no longer be detected.

For years I was told that it was a waste of time to try to figure out what I was reacting to. I was told that there were just too many possibilities happening all at once, so there was no way of knowing for sure. Also, I was probably reacting differently than others. Besides, this wasn't life-threatening so just learn to live with it. I accepted that opinion but always had a nagging doubt, especially after it became too intrusive to just live with it.

If it's true that there are that many causes and effects, then shouldn't someone be looking at what they are? I mean, it's not like trying to find a once-in-a-lifetime event. There are all kinds of things happening all the time. Shouldn't someone at least try to figure out ways to simplify the task? I mean, if sources of exposure are that prevalent, then aren't there enough of us being affected to make a study worthwhile?

It is at this point of differing perceptions and experiences that the complexity increases, triggering fear and other psychological issues. You no longer have clear information that you can trust. The mistrust occurs because of the discrepancy between your personal experience and the authoritative conclusion of the expert. People in this situation often leave their doctor's care to pursue other means of solving their complaint. They become hypervigilant, potentially exposing themselves to even more harm. Doctors and patients both need to realize the complexity and the implications of what is happening.

THE POWER OF THE COMPLAINT EQUATION IS THIS:

1. Its richness allows different individuals to have very different reactions, perhaps unlike anyone else in the world.
2. Its simplicity allows us to begin all journeys of complaint reduction by considering only two factors: susceptibility and exposure.

Chapter Five

Susceptibility and Exposure

Susceptibility and Exposure

Everyone reacts differently to different substances. Each of us has his or her own unique combination of susceptibilities and exposures at various levels.

Some people react to minuscule amounts and others require a "bucket full of poison" before they suspect that they have been harmed. Some people can drink arsenic and ask for a cyanide chaser," while others just enter a room with a closed bottle of perfume and are sick in bed for three days.

Why is it so surprising that we are not all alike? We readily accept that we cannot all run a four-minute mile, or even run a mile. We don't all like the same music, read the same books or watch the same TV programs. We can not even agree on the volume setting for the TV. We sleep differently, eat differently, work differently, play differently. Our experiences and our perceptions of those experiences are as varied as the number of individuals in the world. Despite many commonalities, each of us is different and has a unique set of experiences.

In wildlife, it is not only possible, but common, for animals to react to only a few individual molecules of a substance. Male gypsy moths begin their complex mating behavior upon the detection of a *single* molecule of the female's pheromone. Bloodhounds used for tracking can follow the scent of a kidnaped person from bushes alongside a multi-lane expressway several days after the victim's car has passed by. The chemical added to natural gas to give it its distinctive odor is designed to be detected when only a few molecules are present within a whole house. When diagnostic labs need to increase the sensitivity of their chemical detection instruments, they don't add another piece of mechanical equipment. They add a person. Most humans are many times more sensitive to odor than even the most technologically advanced lab instruments. Yet, some of us cannot even smell spoiled milk.

Whoever thought up the crazy idea that when it comes to "medical" experiences or issues of public health and safety we are all identical? Why do the sharpest medical minds in the world set diagnostic and treatment standards based on the assumption that we all react identically?

Regardless of the motivations, which could fill another book, the fundamental assumption behind this categorical blindness lies with how safety standards are established to begin with. The key concepts are: Public Safety and Statistical Correlation (the bell curve).

PUBLIC SAFETY

Public safety standards are not only essential but have proven highly effective. I have no quarrel with them — as long we remember that these standards apply to the public at large and not to any particular individual. Rarely does any public official in the news even address the point that the specific individuals that make up the public may experience the events much differently than the "average" citizen, whoever that is.

To further confuse this issue, public health officials seemingly have as their prime directive "prevent public panic." Again, preventing panic is important but officials tend to put such a large emphasis on it that the rest of the message gets lost. And it is the "rest of the message" that contains the specific information that is often of most importance to individuals.

Public health and safety is based on the Cause-and-Effect model discussed in the Introduction. This model assumes one cause for each effect in an objectively predictable manner. Any events that don't fit the model are discarded as an anomaly. However, personal experience is exactly that, an anomaly. Plus, many events don't have a single cause and a single effect. What may be more useful in understanding individual events is a model based on Influence-and-Tendencies (as discussed in the Introduction.)

Finally, they often lose sight of the initial problem and ask the wrong research questions. They lose their focus. A typical scenario is when an individual has a child that has been diagnosed with an ailment such as autism or cancer. They discover other children with similar ailments. A "cluster" of illness is perceived and an environmental toxin is suspected. When the cluster becomes large enough, public health authorities become involved. Their objective is to determine if a statistical correlation of cause exists between the environmental exposure and the illness.

One of the first steps is to re-examine the original children to see if they truly suffer from the claimed illnesses. This is only logical and is good science. Usually, most of the children are eliminated as having a similar illness rather than the one being studied. One recent study eliminated about 95 percent of the children. Of the five percent remaining, no correlation was found between environmental exposures and the illness. The conclusion was that there was no scientific evidence, in that cluster, to support the conclusion that environmental exposure was a cause of the illness. And I have no argument with that conclusion.

But what about the 95 percent of the children who didn't have that specific disease? What were they suffering from? Illusion? Hypochondria? Could it be that the research question ought to have been shifted from "does environmental exposure 'X' cause illness 'Y'," to "does environmental exposure 'X' cause any of these 'look alike' illnesses?" Although that is beyond the scope of the original study, isn't that an important question? Just because the majority of the children don't have the illness being investigated by the study doesn't mean they aren't sick. Perhaps their illnesses are precisely the effect of environmental exposures that the researchers originally looked for.

This story is an example of a system of closed feedback loops. A closed system sees only what is in the loop, rejecting all else as an anomaly. What I'm advocating is that the investigators open the loop to allow other research questions that can tell the whole story about the whole group.

THE "BELL CURVE"

The bell curve is the second key concept of determining public safety. Again, the problem is not the bell curve itself, but rather how public authorities misuse the data that generates the curve. We, the public, then assume that their presentation is "gospel" and further perpetuate the distortion by forcing our experiences into the framework of their statistical distribution curves.

Statistical distribution curves can lead to distortions of information and of understanding. We are most familiar with the bell curve from school. Prior to the use of the pass-fail school grading system, the debate about grading methods was whether or not students should be graded "strictly by the curve" or by a "modified curve."

"Strictly by the curve" meant that there must be a total distribution of all grades of "F" through "A" throughout the group. The lowest score was an "F," whether it was a 30 or a 90. The highest score was an "A," whether it was a 90 or 30. The rest were distributed in between.

BELL CURVE FOR SCHOOL GRADES

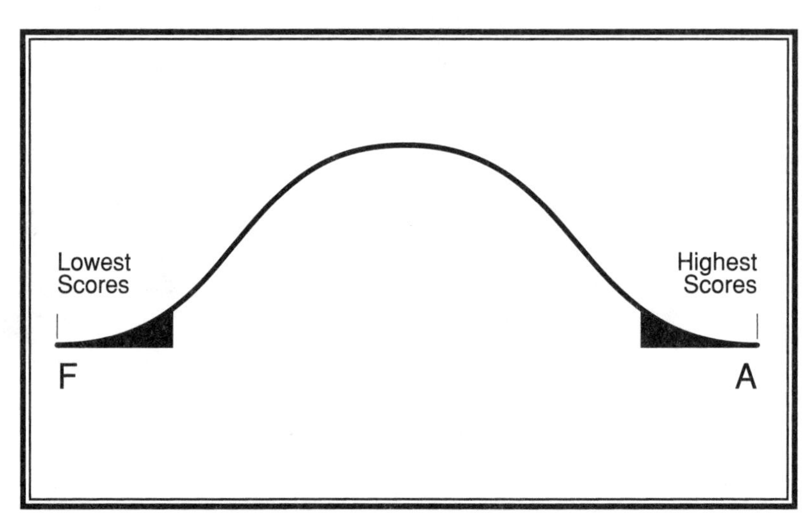

Grading by the curve means that the lowest grade is the beginning of the bell curve. Regardless of what that score is, there will always be a lowest score that will receive the lowest grade. That is what defines a failing grade. If the failures persist the student eventually quits or is dismissed. He or she is no longer included in that group and is eventually forgotten. They are treated as if they no longer exist. But the truth is that they do.

"Strictly by the curve" meant that there would always be at least one "F." Someone ***had*** to fail — by definition. In fact, the whole curve depended on that failure. Without it, the curve had no beginning. With no beginning it could have no ending. One of the realizations that led to the "modified curve" was that whole groups of students were being stigmatized as failures, when in fact they had learned substantially and actually became productive citizens. The consequences to the people being graded eventually became as important as the integrity of the grading system itself.

Public health and safety standards are largely determined in a manner similar to grading "strictly by the curve." All studies begin with the first occurrence of harm, whether it is a test for cancer, toxins, infectious disease, or physical harm. The first occurrence starts the bell curve. The "last one standing," so to speak, determines the end of the curve.

BELL CURVE FOR TOXICITY

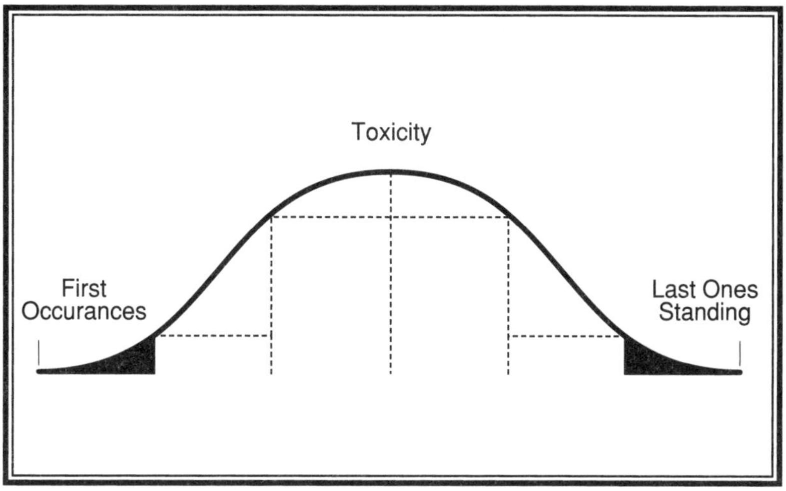

Toxicity

First Occurances

Last Ones Standing

The bell curve is also used to measure toxicity and other forms of potential harm. And just like with grading students, the curve begins with the first occurrence of whatever is being measured and ends with the "last one standing."

The next step is to use the curve to help determine health and safety standards. This process is heavily influenced — and naturally so — by society's opinions of cost vs. benefit along with the inevitable political maneuvering.

Then testing methods must be developed to determine violations of those standards followed by medical diagnostics to determine harm.

But what if the alleged event isn't included in the catalog of standards? Should that mean we need to consider the new data and revise the standards as needed? Or should it mean that the only events we accept as "real" are what are already defined?

Rather than primarily protecting the public, the main result of public standards seems to be to grant the right to sue and to increase the likelihood of prevailing if those standards are violated. All else is often dismissed as incidents of personal illusion.

All those at the far right end of the curve are naturally resistant. They are the ones who can "drink arsenic and ask for a cyanide chaser." We don't have to worry much about them. They probably don't even need safety standards.

However, those at the very beginning of the curve are the ones who are sick in bed for three days from just a closed bottle of perfume somewhere in the house. They are so reactive that none of the established safety standards will protect them.

And therein lies the problem. No standards apply to them. They could, but they aren't intended to. These individuals aren't protected because they are being "graded strictly by the curve." Society seems to need this hypersensitive group — these "necessary victims" — in order to start the curve to maintain the integrity of the system. And as a society and as individuals, we also need them in order to feel safe. If we know that the danger is harming "them," but not "us," then our sense of safety and well-being seems more secure.

But for those that perceive they *are* being harmed, their sense of safety and well-being does not seem more secure — it feels violated! And the only others they can compare their situation to are the "safe" ones.

WHICH BELL CURVE?

Until 1989, state-of-the-art toxicity and safety information was based on ***the*** bell curve. The curve started with the first occurrence of what was being measured. But no one asked what happened prior to the first occurrence.

It was a lot like the ancient mariner maps in old geography books. They showed Europe with Great Britain to the west. Beyond that was water. Beyond that was the warning "There Be Monsters." And no one explored those regions.

It was a classic case of categorical denial, a form of closed system. Rather than conduct an exploration to see what was actually out there, they assumed their biases and fears were true. And if they were wrong? We can imagine their response as, "Well, obviously there could be nothing of any relevance or value. Investigation would be a harmless waste of time. Or at the least, a personal amusement!"

In terms of toxicity and safety, all that changed in 1989. That was the year that Drs. Nicholas Ashford and Claudia Miller, commissioned by the New Jersey Department of Health, published their groundbreaking report, *Chemical Exposure: Low Levels and High Risk.* They looked beyond the earlier assumptions and asked a fundamental question, "What occurs at exposure levels below toxicity?" Among several key findings were these:

1. As the next diagram clearly illustrates, they discovered that a lot happens prior to the toxicity curve. That's where the atopy and the sensitivity curves are located.

THREE BELL CURVE RELATIONSHIP

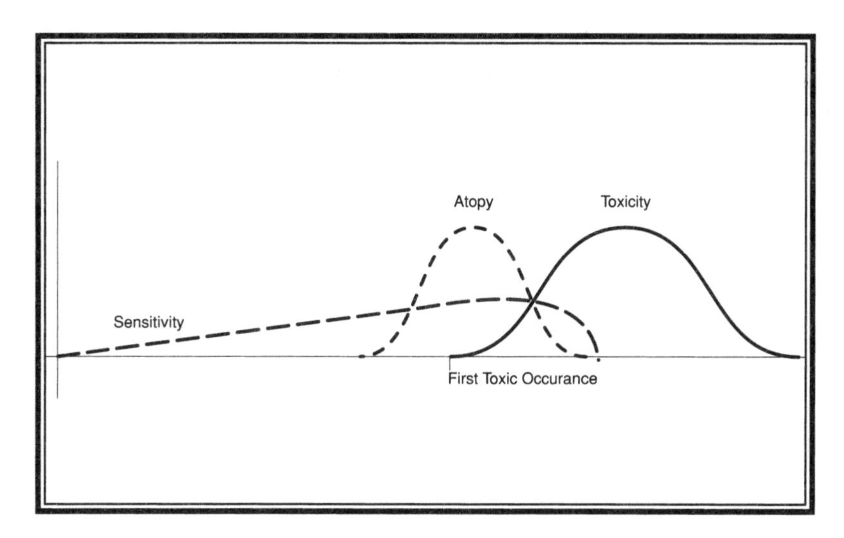

In 1989 Drs. Nicholas Ashford and Claudia Miller asked a very fundamental question: What happens prior to the "first occurrence"?

As the curves above show, a lot happens. In cases of exposure to chemical toxins, they measured two phenomena in addition and prior to toxicity: Atopy and sensitivity.

This is a classic example of how the question often determines the answer. Change the question — change your point of view — and new data is often revealed.

Adapted from *Chemical Exposures*

The atopy curve begins abruptly and ends abruptly. Whatever percentage of the group is going to react will do so nearly at the same time and the same levels, but *prior* to the beginning of the toxicity curve. Also notice that it overlaps the toxicity curve. In that area of overlap, it is impossible to differentiate between atopy and toxicity.

The sensitivity curve also begins prior to the toxicity curve and even before the atopy curve. It begins near zero. It then rises only slightly and very gradually. Unlike the atopy curve, where whoever is going to react does so quickly, few people react in the sensitivity curve range. When they do, the same person — because of many subtle variables — can react at different levels at different testing times. Notice too that the sensitivity curve also overlaps the toxicity curve and the atopy curve. In the area of overlap, it is impossible to differentiate between the three types of reaction.

LONG TERM VS TOXIC

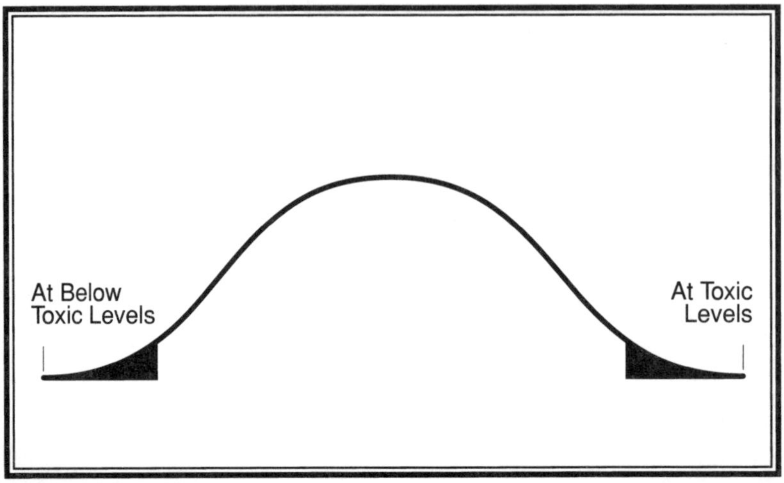

Long-term exposure to low levels can often be more harmful than toxic exposures.

Toxic exposures, by definition, are harmful. In many cases the harm is perceived as pain of some sort, pain that is usually strong enough that you immediately move away from the source. The exposure level is high so the exposure time is short.

But if exposure to a toxin cannot be personally perceived, as with carbon monoxide, for example, the person being exposed is still in grave and immediate danger and the symptoms will be fairly rapid and unmistakable. This is a situation where test instruments are critical for preventing harm by keeping the exposure time short.

But what if exposure to a toxin is below the defined level of toxic harm? Does that mean that "nothing" happens? Perhaps it just takes longer for the symptoms to manifest. Or perhaps the effect is to slowly weaken the immune defenses, allowing an opportunistic infection, for example. The key here is that neither test instruments nor the human body generates the warning information. So the person stays and continues receiving the exposure.

A further complication is that recent research shows that exposure to low levels of some chemicals may not generate physical changes in the body but may generate physical changes in the emotional control centers of the brain, changing mood and behavior. Also, some chemicals that are nontoxic are suspected of emulating hormones in the body, affecting the endocrine system.

2. Long-term exposure to low levels of chemicals are often more harmful than a toxic exposure. While that may seem surprising, consider that a toxic exposure is, by definition, so strong that neither the human body nor the mind can deny the occurrence of that event. Action to stop or avoid the exposure is usually extremely rapid. However, if the exposure is not extreme

enough to penetrate the normal denial response, the exposure can continue to occur for prolonged periods of time. There is little research data about exposures lower than toxic levels for the simple reason that before Ashford and Miller, there was no reason to explore. It was assumed that there was nothing to be found.

3. There is considerable variability of reactions both within the group and by individuals to a wide range of exposure levels, particularly for the sensitivity curve data. The slope of the line is so flat that it can take a comparatively large increase in exposure to increase the number of people who react. Likewise, because to the variability of susceptibility of individuals — as in you caught a cold this week but not last week because your resistance decreased — the same individual may react at widely varying levels at different times.

4. Reports from workers in at-risk environments are not a reliable measure of safety or lack of harm. For example, people who work around petroleum distillates, pesticides or other controlled sources typically claim that they haven't been harmed in all their years of exposure. Setting aside the accuracy of that statement for the time being, it is very likely that they have a high resistance to the high ambient exposures. If they didn't they would either be on disability or have left for another job. In other words, some sort of natural selection process seems to be operational for workers in at-risk environments. Those who stay are, of course, resistant. Those who are susceptible, perceived or not, either get sick and leave or they just aren't comfortable. They prefer something else. Either way they seek work elsewhere, leaving behind a selected group of highly resistant workers.

IMPLICATIONS

These findings raise several implications, some rather surprising:

The mere use of the bell curve confirms the variability of both susceptibility and exposure. If there is not a variety of reaction responses to various levels of exposure among individuals, then there is no "curve" to the bell. It would be a straight line, running from top to bottom. Everybody would react at exactly the same point of exposure. There would be no "first to fail" or "last one standing." They would all be harmed at the same time. But because the curve is bell shaped, you *must* react to an exposure differently than at least some of the others.

The bell curve describes a group. It says absolutely nothing about any specific individual, not even those who made up the group. *The curve says nothing specific about you.*

If you are at the very beginning of the curve, then you, by definition, have been harmed. You are a victim created by the very system that was designed to protect the public. But this system was not designed to protect *you* or any other specific individual. It was designed to protect the statistically calculated masses.

Before anyone, regardless of their credentials, can make a claim as to whether or not you may be harmed at any specified level of exposure, they must first determine *where you are on the curve*. The further to the right you are, the less likely that you will be harmed. The further to the left you are, the more likely that you will be harmed. If they don't know where you are on the curve, they have no idea whether you have been harmed or not.

Before anyone, regardless of their credentials, can make a claim as to whether or not you can be harmed at any level of exposure, they must first determine *which of the three curves* you are on. Are they talking about the toxicity curve, the atopy curve, or the sensitivity curve? Then, as above, where on that curve are you?

Before anyone, regardless of their credentials, can make a claim as to whether or not your exposure experience is *a mental disorder*, they must first place you precisely on one of the curves. If they can't, then you may be at the extreme left side of any of the curves and be experiencing a legitimate physical reaction to an actual physical substance. If they don't know, then they cannot exclude a physical cause. Any diagnosis of a mental disorder is by a process of exclusion that first excludes *all* physical causes. If it cannot, then the possibility exists that the cause is physical.

A *mistaken diagnosis of hypochondria* and other related mental disorders usually *causes more harm* to the individual than does the exposure itself. Dr. Henry Vyner, author of the book *Invisible Trauma*, concludes that less harm occurs to the patient when the diagnosis errs toward the side of physical causes rather than to hypochondria. Or in simpler terms, rather than assuming that the proper action is "patient, heal thyself," authorities ought to instead "First, do no harm." Therefore, for your own mental health, don't assume that it's "all in your head." You may very well be one of a very few who are aware of what is happening.

Test instruments, as to what substances they can detect and at what levels they can detect them, are based on public health and safety regulations. They are not designed to detect "everything" at all levels. You may very well be reactive to something that the instrument is not designed to detect, such as common allergens. And if you are a hypersensitive individual, then the instruments may not be sensitive enough to detect anything at the atopy and sensitivity levels. Therefore, just because a test instrument does not detect "anything" doesn't mean there is nothing to be detected or to be exposed to. Nor does it mean that you are *not* being exposed to something you are susceptible to. These may simply be false-negative results. Also, test instruments cannot by themselves determine what is wrong. Test instruments can only provide data that must then be interpreted. The interpretation is based on what question is being asked. Test instrument data is a means to an end, not the end itself.

Laws and regulations for public health and safety are based on the bell curve and are designed to protect the majority of the general public. Their fundamental purpose is to establish a right to sue if an exposure can be measured as a violation of those regulations. Regulatory compliance does a terrific job of improving the safety of the public, as a whole, from a few specific substances. But it is not much help to individuals, especially the hypersensitive ones. Public health and safety laws are a necessary starting point, *but if you are more susceptible than most people, do not rely* ***solely*** *on public standards for your* ***personal*** *safety.*

SOURCES OF AUTHORITY

So, if none of these authoritative sources is of much help, where does the specific information come from? It has been and continues to be generated by the experiences of people like you — the ones who have already begun their own personal journeys.

There is a small percentage of people, whose numbers are, unfortunately, increasing far too rapidly, who have had exposure problems for years. They are extra-sensitive to all kinds of exposures, reacting long before most of us are even aware of them. These are the people at the far left of the bell curve, even to the far left of the sensitivity curve.

They have had to struggle, often blindly, with desperately having to do *something*, but not having sufficient information to make low-risk decisions. They have had to "take the leap" and act without knowing what the outcome would be. They have made many mistakes but have also had many successes. Using little more than trial-and-error techniques, they continue to learn what works and what doesn't.

One implication of their experience concerns the meaning of "safe." Rather than define "safe" in terms of the statistical distribution of the general population or in terms of regulatory law, they define it as what is "safe" for hyper-reactive people. If they can find products, services, procedures and conditions that are "safe" for most of them, then the risk to society in general is also greatly reduced. In fact, their experience may be a more reliable standard of *individual* safety than the politically tinged *public* standards based upon group studies and the calculated averages of, as the old saying goes, "lies, damn lies, and statistics."

Although their methods of trial-and-error and their level of understanding appear to be rather crude and not very scientific — anecdotal, even — the results are often stunningly effective.

An historical example occurred in London in 1849. Dr. John Snow published a pamphlet entitled *On the Mode of Communication of Cholera.* Dr. Snow claimed that the disease was being spread by contaminated drinking water. He further observed that most of the cholera victims in a particular housing area were confined to one building and they got their water from the Broad Street pump. Occupants of the other houses in the area, however, had a different well and remained relatively free of the disease. The Broad Street pump was located within a few feet of a human waste drain pipe coming from the house with the epidemic. Dr. Snow removed the handle from the Broad Street pump and the epidemic ended.

Critics of his action and conclusion could claim, quite correctly, that short of witchcraft, a pump handle could not cause any physical problems other than a bump on the head. They could attribute his removing the pump handle as coincidental to the end of the epidemic. What proof did he have? Where were the double-blind, controlled studies?

What they would be completely missing is that the pump handle was not the issue. The pump handle neither caused the epidemic, nor did it stop the epidemic. The *removal* of the pump handle stopped a *sequence* of events. It effectively isolated a source of disease from the housing occupants and that is what allowed the epidemic to run its course and eventually end.

How could a pump, such as the one above, cause or prevent an infectious epidemic? It can't, but that is asking the wrong question. A better question is, what influence does the function of the pump have on the sequence of events? If there is one, then what change can be made to influence a different sequence?

Contemporary examples abound. It is not unusual for the highly susceptible individual to be the first to smell a gas leak at work, for example. Even though their warnings are often ignored initially, they do provide an early warning system of sorts. This is another example where the Influence-and-Tendencies model is more useful than the Cause-and-Effect model.

A common way of describing these hyper-aware individuals is the overused "canary in the coal mine" analogy. This is derived from the practice of coal miners near the turn of the century. Periodically miners would die in the mines. They had no idea why this happened. They had even fewer ideas on how to stop whatever was causing these disasters.

If the canary dies, no one goes into the mine shaft. If the "necessary victim" survives, then the miners go to work. But if a more-sensitive person gets sick from low levels of a toxic gas, he is often ignored or accused of being lazy or a hypochondriac. He will either get sick or eventually leave that job. Meanwhile, the others scoff at the idea that the job can be dangerous to health. They have worked for years without any problems! A sort of covert "natural selection" process.

However, they cleverly developed an early-warning system. Rather than risking their own lives, they would first lower a caged canary down the shaft. If the canary lived, they knew it was safe for them. If the canary died, the miners didn't work. These canaries became a "necessary victim." But at least the miners acted on the results of their test and didn't blame the canary's calamity on a psychosomatic illness!

A reversal of the "canary" story was told to me by a client. Her observation was that when exposures at work began causing problems with her and some other employees, management's course of action took a sinister, but all too common twist. Instead of recognizing the complaining "canaries" as harbingers of potential danger to others, they pressured the complaining workers to quit. They replaced them with people who didn't complain. These new employees may have been affected and aware of the exposures, but they wouldn't complain because they knew what happened to the previous employees.

Rather than heed the clues from their canaries, management replaced them with different canaries who gave no warning! That's like removing the batteries from your smoke detector because you don't want it disturbing your sleep.

Standards of health and safety are essential for the public. And they are an essential starting point for any individual. But public standards do not address all areas of susceptibility and exposure. Also, they are not specifically applicable to an individual.

This point is becoming increasingly important as the size of the group of people being harmed by toxic chemicals, sick buildings, common allergens, respiratory irritants and asthma-inducing triggers — those near the beginning of the bell curve — expands beyond the boundaries of contemporary determinations of the bell curve.

Estimates from health organizations and government agencies range from 30-60 percent of office workers complaining of being ill because of sick building syndrome (SBS). These estimates do not include allergies, respiratory irritants, asthma triggers or individual complaints. They are based on group data when more than 20 percent of the occupants are uncomfortable enough to complain.

This situation signals to me that "something" may be increasing either the exposure levels or the susceptibility, or both, of the majority of the population. If true, then the increasing number of complaints are no longer limited to the statistically calculated 2% category of "necessary victims." And that would challenge the credibility of the very standards that have previously held the public's trust. If so, then society's starting point for health and safety may need to change.

Chapter Six

Starting Point Difficulties

Starting Point Difficulties

The general plan and the execution of the steps in Chapter 2 is relatively easy. What is difficult is the application of the plan and the evaluation of the results. Because your starting point often dictates where you end up, it is critical to explore those difficulties so as to avoid unnecessary waste of time, effort, and money.

The difficulties begin not with your lack of housecleaning skills. Nor are they due to some occult force that intervenes in your life or some defect in your moral character. Rather, the initial failure of this simple plan lies in definitions and assumed meanings of common words and procedures. Although the following information may seem trivial at first, if you experience difficulty in safely reaching your destination of a complaint-free indoor habitat, this will be an important section to review.

Successful starting points do not rely on some expert telling you what they are, and then believing with all your heart that his guidance will safely transport you to the final destination. Starting points are not just the process of deciding on a goal, writing it down, and then trusting in divine intervention to supply the reward. Rarely are they what family and friends say they should be. Finally, successful starting points are not simply just "taking off" from where you are.

The greatest difficulty in determining starting points is how you get the information you need when you need it. Whom do you rely on? If those you ask don't truly understand the *individual* nature of the journey you are about to undertake, how can they accurately advise you? To better understand the process, compare this journey to a geographical one.

A true starting point for any process is to first determine, as best you can, where you actually are. If you don't first mark your location on whatever map you are using, you have no idea where you are or how to get to some other destination.

For example, if you want to get to Boston, a map will show you the way. But first you must know where you are. Getting to Boston from Dallas will be much different than getting there from Moscow. If you don't know where you are, you can't figure out how to get anywhere. You will just roam around aimlessly until you get to "someplace" else. And because you don't know where your new location is, you won't know how to return to where you came from.

> *If you don't specify which complaints occur with which indoor exposures, then how will you know that the proper sources have been removed? If your first attempts at source removal are only partially successful, how will you determine what worked and what didn't so that you can begin again?*

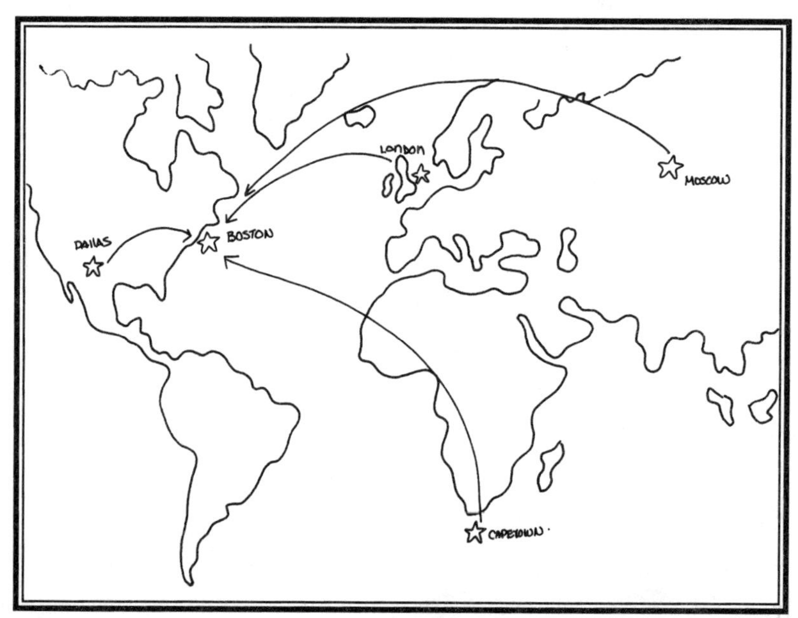

The specific instructions for how you travel from one place to another depends on where you start. Going from Moscow to Boston will require very different methods and techniques from those used to travel from Dallas to Boston. And if you don't know the subtleties between an airplane, a boat and a car, you will probably experience some difficulty on your journeys.

You also need to know the geographical features, the "lay of the land." Your friends' ecstatic tales of their latest Mediterranean cruise may have been so compelling that you'd like to take a cruise, too. But instead of going to the Mediterranean like your friends did, you want to take a cruise from

Dallas to Boston. It can't be done. But if you want to leave from London and go to Boston, you could make that trip easily.

Likewise, you may want to get rid of your exposure to cat dander by vacuuming the carpet, not realizing that the vacuum cleaner may actually be blowing the dander in the carpet back into the air and allowing it to then accumulate in numerous other locations.

Next, you will need to evaluate your traveling style. Do you prefer luxury or bare bones? Slow and scenic or fast and direct? What will your budget allow? Do you have to get to Boston for a specific meeting or can you arrive whenever you desire? Are you traveling alone or with a family?

In terms of cleaning up your indoor environment, do you want an absolutely reaction-free house or just enough improvement to make the symptoms tolerable? What will your budget allow? Is there a timetable that must be met? Must the results satisfy only yourself or must they also satisfy other family members?

The answers to the travel questions may seem trivial because you already have travel experiences. The process for making decisions is familiar, and the consequences of mistakes aren't usually life-threatening. Besides, reliable maps, guides, and modes of transportation are readily available. What you don't already know you can obtain from known, reliable sources. It's no big deal.

But how do you make decisions if you have no direct travel experience? Whom do you rely on for guidance? The telemarketing phone call offering the trip of a lifetime for an unbelievable low price? Or the advice of trusted family and friends and their referrals to travel agents they have successfully used in the past?

Now, suppose you need to travel to a location that has no reliable means of transportation for getting there. And furthermore, the journey is so infrequently taken that there are virtually no maps available. Would you even attempt such a trip?

Imagine, for a moment, that you either *have* to make the journey or you suddenly realize that you have already begun and can't find your way back home. You are then forced to continue without reliable knowledge and trustworthy sources of information. Your journey may begin to acquire the characteristics of a mythical quest, complete with archetypal issues of survival, identity and personal transformation.

Now, suppose the year is 1698. How do you get from Dallas to anywhere else when Dallas doesn't yet exist. Actually, the physical location exists, but it isn't *called* Dallas. So if you ask someone how to get from Dallas to Boston they won't even know what you are talking about. But if you can describe your present location by its geographical characteristics or by comparing it to nearby locations, you will eventually find someone who understands where you are and what you need. Then they can tell you how to get to where you want to go.

How can anyone else guide your cleanup efforts if the language and the tools for that activity don't exist yet?

Even if someone does understand what you want to do, they may not give you reliable advice. They may lie to you about how to get to Boston as a way to convince you to buy their supplies. Or they may actually take you to Boston themselves, but at an outrageous cost. And because you've never been to Boston before, how will you know you've actually completed the journey? You may actually be in St. Louis, only partway to your intended destination.

Some companies understand that you want to remove mold from your house, but they insist that what you really want is to have them clean your carpets, claiming that carpets are the main location for mold. Or they may actually suggest the correct action but only accomplish part of what you need done. How will you know they actually removed the mold from your house and not just from your carpet?

If you are following someone else's instructions, your starting point will actually *determine* your destination. For example, if the directions given to you are to travel 250 miles east, then 90 miles south, you could end up anywhere. If you start in Boston, your destination will be somewhere in the Atlantic Ocean. But if your starting point is San Francisco, you will go to the desert.

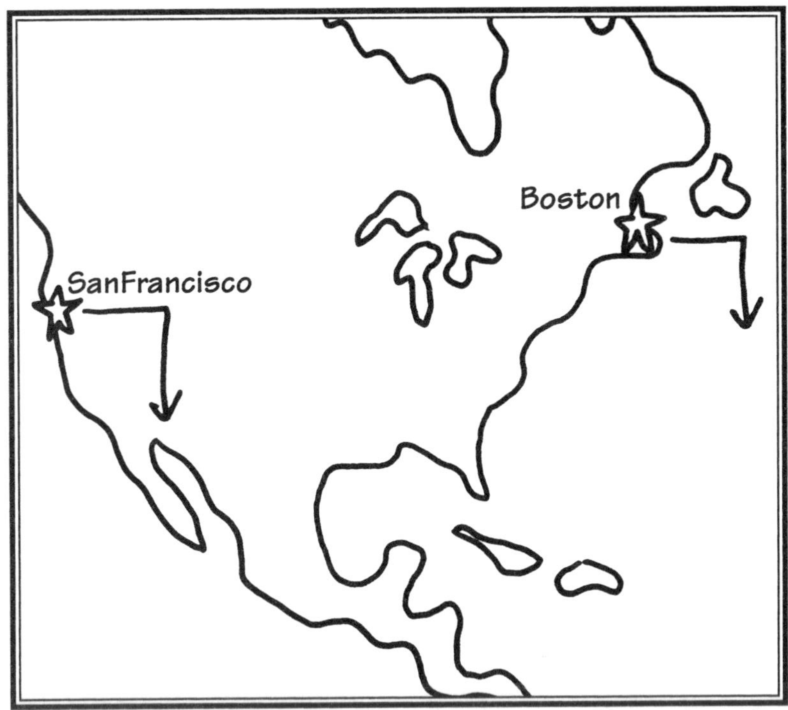

Blindly following instructions can create serious problems, depending on your starting point. If you prepare for a desert destination but leave from Boston, you won't achieve your desired results.

If the instructions you are following are to vacuum all carpets every day, you could end up with any kind of results. If you have asthma attacks caused by dust mites, then vacuuming carpets will pick up the dust mite allergen and blow it back into the air, increasing your exposure. Or the dust mites may not even be in the carpet, but in the mattress. If your starting point is to stop heavy coughing or to avoid a headache when you are exposed to perfume, vacuuming the carpets will have no effect at all. A simple instruction might have numerous variables and a multitude of results.

Even if you have an accurate geographical starting point and know your destination, you still have to understand why you are going on your journey. For example, if you want to go to Boston to see the theater performance "Phantom of the Opera," you first need to know if it is playing in Boston. Perhaps it has moved to Toronto or to Houston. If so, going to Boston will only result in disappointment, frustration and unnecessary costs.

The starting point may be the recent news reports that cockroach allergen is responsible for most asthma attacks. Therefore, the purpose of removing cockroach allergen is to stop your asthma attacks at home. But first you need to know if there is cockroach allergen in your home. Perhaps dust mites are the cause of your asthma attacks. Cleaning up cockroach allergen will only result in disappointment, frustration, and an unnecessary financial burden.

Finally, how you evaluate the results of your journey also is critical. If you use the location of the show "Phantom of the Opera" as your landmark, you may travel around the country until you find that show, even if it isn't in Boston. However, you would declare that you are in Boston, only to discover later that you actually are watching a community playhouse production of "The Phantom" in Blaine, Missouri.

How you evaluate the results of your house cleanup is critical. If you use the removal of cockroach allergen as your landmark, then you may spend hundreds of dollars for carpet cleaning, duct cleaning and a fumigation. Laboratory tests prove that there is not even a trace of cockroach allergen anywhere in your house. You would then declare that you have had a successful cleanup only to discover later that you actually are reacting to dust mites in your bedding.

THE EXPERIENCE OF THE JOURNEY

Establishing your starting point and its associated characteristics is only part of your traveling experience. The greater challenge is the actual experience of the journey itself.

Preparing for an uncharted exploration can be compared to the adventures of Lewis and Clark, for example, or other early pioneers. When they first started their journeys, what few maps they had guided them only part of the way. They had to go beyond the boundaries of known information and generate new maps as they went.

Not only was exploring without a complete map difficult, but only very peculiar personalities tended to undertake such adventures. Early pioneers often were seen by society as undesirables. Some later became invisible to society and were often completely forgotten. Some of them weren't even missed.

Travel can be easy when you have a known, dependable means of transportation, over familiar terrain and with a defined destination in sight.

The isolation and other hardships of their travels were tremendous. Many of them didn't survive their ordeal because healthful food and adequate shelter were hard to come by. Other humans they encountered were frequently hostile and sometimes deadly. Expert medical care was nonexistent. Because they had no external support system, they had to create their own internal one.

The few who did survive are now viewed as heroes and are celebrated for their advancement of knowledge about our world and the people in it. Although specific details, and sometimes even their general conclusions, were later proven to be inaccurate, like Christopher Columbus thinking he was in India when in fact he was in the Caribbean, enough information was accurate to later guide others. Corrections to the course were made by the next waves of explorers. Even today, our maps are constantly being corrected and refined as new technology and needs develop and as political upheaval causes factual changes.

Our contemporary journey within our own indoor human habitat is not much different from that of early pioneers. We have few reliable maps that apply to our situation. And the few that do only take us a short distance from where we start. We aren't even sure where we are going or how to know when we've arrived. We don't know beforehand if we can get our life back the way it used to be — if we can roll our snowball back to the top of the mountain where it all began.

Drifting along with no visible means of support, no clear point of departure and no destination in mind can leave one quite disoriented and lost.

The major differences between geographic explorers and victims of indoor environmental exposures are that ***our terrain is invisible rather than geographic and we rarely know our starting point because we did not begin our journey by choice.*** We didn't even know we had started a journey until we were well on our way. In other words, we begin our journey by being lost.

Contributing to this sense of being lost and alone is the disagreement about whether our journey is necessary or that our destination even exists. While many of us find our experiences undeniable and rich in detail, some of the greatest experts and authorities insist that the experience exists only in our heads, that it is a hysterical manifestation of our imagination.

These experts insist that no scientific evidence conclusively proves the actuality of what we experience. They insist, in essence, that the world is flat and we will only destroy ourselves and unnecessarily harm our loved ones by "sailing over the edge of the horizon." Therefore, they continue, there is no need to make a new map. The hundreds of maps already in existence will do just fine. There is nothing more of importance to discover.

Their *public* maps may be adequate for them and even for the majority of people. However, *your* experiences have proven that public maps are inadequate for navigating *your* life. Because their maps don't safely guide you and because you have to either act on your own or surrender to being victimized, you are going to have to create your own uniquely individual map.

The first step, which you took at the beginning of Chapter 2, is to describe as best you can where you *think* you are. As you generate new information about how your complaints change as you change your exposures, you will create a series of new starting points. By continually and more expertly redefining where you are, you can actually wander around and still establish an accurate and useable starting point. In fact, the more you do wander around at the beginning, the more information you will generate, the more detailed your map will become, and the more likely you will increase your chances of success.

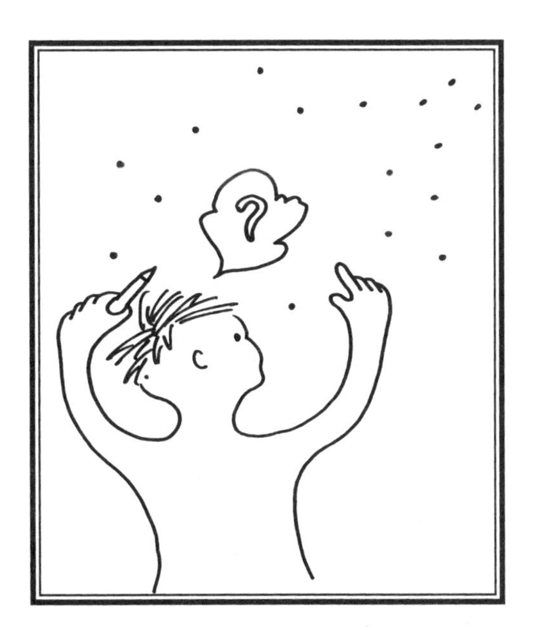

An effective plan with appropriate tools and a clear understanding of relevant concepts still requires a specific starting point. Once your starting point is defined and the sequence of dots has been established, you will "get the picture." Otherwise, you may easily become lost and confused.

You can then compare your map with those of others you encounter along the way. With experience, you will be able to travel from "wherever you are" to "wherever" you choose to go. You then have better information with which to make better choices to stop your complaints, prevent future complaints or even to *enhance* your indoor habitat.

Chapter Seven

Personal Impact Rating (PIR)

Personal Impact Rating (PIR)

Susceptibility and exposure are both necessary for a complaint to occur, according to the *Complaint Equation*. Break the connection between them and you have no problem. But in complex situations where a variety of actions are necessary, some of which are expensive, require introducing new sources of exposure, and are disruptive to employment and lifestyle. It would be helpful to have a way of prioritizing both the need and the appropriate caution.

For example, is your complaint with dust mites a minor nuisance or does it send you to the emergency room with an asthma attack? How does it compare to your complaint with perfume, cigarette smoke, new carpet or tree pollen? Can you hold a half dozen cats in your arms with no problem, but not be within 10 feet of someone who has a cockatiel at home?

You need to fine-tune your evaluation of the situation. You need a way to rate the *impact* of the complaint equation to provide guidance about what is necessary, determine the sequence of steps, what is unnecessary, and what to avoid. You need to refine your starting point by being clear about the type of impact a particular exposure does or does not have on your life.

It would also be helpful to have a comparative context for your situation. Complaints without a realistic context have a way of expanding to the farthest borders of your concerns and fears. Comparative boundaries help to realistically assess the level of danger or comfort that may be inherent in your particular situation. This is not unlike the need for society to establish its "necessary victims." The difference for you, though, is that you don't need to dismiss those less fortunate than you as blameworthy. You only need to know how your situation compares to others for the purpose of making better choices about what to do.

Your ***Personal Impact Rating*** **(PIR)** will be your main guide for determining the extent of the effort and caution required to stop your complaints, prevent new exposures and compare your situation to others. The higher your rating, the more diligent and concerned you must be. The lower your rating, the more carefree you can be.

The PIR includes six levels of impact from exposures. The higher the number, the greater the impact.

PIR 1 - None, or only occasionally noticeable, like a fleeting thought. This rating won't apply to most readers unless they are interested in preventive measures or want to better understand what their loved ones are experiencing. However, you can have a very high overall PIR but not be affected by certain specific sources.

PIR 2 - Slight - You occasionally notice exposures and it can be ***uncomfortable***. This infrequent, nuisance-type impact does not cause you to stop your routine activities. In the rare event that it does interfere, you may need to take an antihistamine, open a window, or take a shower when you get home. Even under the worst of circumstances, you can delay your response until a more convenient time or place.

PIR 3 - Moderate - Reaction to exposures ***interfere*** with your life. They don't stop you from doing what you want, when you want, but you may have to pause to respond to them by taking an antihistamine, for example, or by using your inhaler. Even though you may have to wait a few minutes before you can continue, you don't have to eliminate that activity from your life.

PIR 4 - Severe - Your complaints become so strong that they are ***intrusive.*** You have to exclude some activities from your life and you frequently must wait until you feel better before you can participate in others. You have difficulty with spur-of-the-moment activities. You can still work, but you are usually very tired by the end of the day. By week's end, your exhaustion is so great that you typically need the whole weekend or longer to recover.

PIR 5 - Disabling - This is when the exposure complaints become so strong and frequent that you have to spend most of your day just taking care of yourself. You may unable to work and may not be able to keep up with routine housekeeping chores. In severe cases, you can no longer function unassisted in many of the basic activities of life.

PIR 6 - Dispossessed - This is an extreme occurrence — an anomaly among the anomalies — that only a few people ever experience. However, it does exist. These are people so severely impacted by such extremely low levels of exposure to just about everything that they can find comfort only under the most pristine conditions. Many live in isolation in the desert of the southwestern U.S. in porcelain trailers. While this impact level may never apply to you (hopefully!) it is important to know that it exists so you can accurately determine your location between the extremes of no impact to total impact. For an accurate understanding of this category, I recommend the haunting book by Rhonda Zwillinger, **The Dispossessed:** *Living with Multiple Chemical Sensitivities,* 1998, published by The Dispossessed Project, PO Box 402, Paulden, AZ 86334-0402. E-mail is rzdisp@northlink.com.

PIR can now help us more fully understand the Complaint Equation. Instead of *Exposure x Susceptibility = Complaint*, it can now be modified as:

$$(\text{Exposure} \times \text{Susceptibility})^{\text{PIR}} = \text{Complaint}$$

The simple change* in the equation illustrates that the higher the PIR the more drastically it increases the severity of the complaint. Likewise, a very low PIR has little to no effect.

To begin to determine your PIR, start with your PIR for your overall condition, taking all factors into account. Later, you can refine it to evaluate the impact of different exposures. (Three general categories of exposure sources are described in the next chapter.) Your overall PIR may be a 4 (severe), but you notice that you don't have any complaints when you are exposed to cats, for example. That observation gives you important information. You know that you don't have to avoid exposure to cat dander and can devote your time and energy to other potential sources.

---

*The complaint equation is not intended to represent an absolute mathematical formula. Rather the intent is to provide an illustration of how the various factors in the equation relate to and influence each other. The original equation represents a progression that most closely resembles a standard multiplication series. For example, if the complaint value begins as 2 and then increases by increments of 2, the series will be 2, 4, 6, 8, 10, 12. The new equation, however, represents an exponential series. For example, if the complaint value remains 2 but is raised to the power of the value of the PIR as it increases from 1 thorough 6, the series will be 2, 4, 8, 16, 32, 64. In other words, low values of PIR have little effect on the total complaint. But as the values of PIR increases, it quickly becomes the dominant factor.

The most common difficulties and confusions occur when you have a PIR of 4 or 5 for several categories. For example, if you are highly susceptible to mold and to chemical odors, you may not want to spray a fungicide to kill the mold. You may only replace the original problem with a different one. Or you may try to find one you don't react to. However, if fungicides have no noticeable impact on you, you may decide to cautiously use them anyway.

Chances are, you will only need this elementary level of fine-tuning your PIR to successfully create your map. But if your map isn't detailed enough for your circumstances, Chapter 13 on *"Stealth" Impact* will provide more assistance.

An additional value of determining your PIR is to reveal those areas where you may want to research and generate more detailed information, personalize the meaning and usability of what you do know, and focus on positive actions for reducing or eliminating your complaints.

CHAPTER EIGHT

SOURCES & THEIR REMOVAL

Sources & Their Removal

Sources and their removal is to indoor exposures what the road symbols and mileage points are to a map. This is the information about the terrain you are traveling and how to actually get from one point on the map to another.

Just what, specifically, are the substances to which you are susceptible and to which you are being exposed? Which ones do you need to be concerned about? Where are they? Where do they come from? How do you get rid of them?

Allergens, irritants and toxins are literally everywhere. There are millions and millions of mold organisms. There are more cats and dogs than there are people. We are surrounded by chemical factories. Sometimes common fragrances trigger an asthma attack. Dust mites lurk everywhere, just waiting to sink their vampire-like fangs into your sleeping body. (This is a myth. They don't bite. It is their feces that cause the problem. Aren't you relieved by that knowledge?) And you never see just one cockroach.

The multiplicity of possible sources can seem so prevalent and overwhelming that many well-meaning people insist that you can't do anything about it. The only obvious response, they add, is to just learn to live with your discomfort. This is fine if it works. But if your "discomfort" becomes disruptive to your life, then a better understanding of sources and their removal may help you stop your complaints.

The magnitude of source removal information also can seem so overwhelming and contradictory that many people stop their efforts out of sheer frustration. But, again, if your discomfort increases to the point that you can no longer ignore it, you need a way to make decisions about how to stop your complaints. Because most information comes from public safety sources or from businesses that want to sell you their product or service, it may not be as reliable for your situation as you need it to be. You need to understand the fundamentals of sources and how they can be removed, isolated or reduced so that you can generate your own accurate information to help you with your decisions. It will also be indispensable for analyzing what went wrong if your complaints increase instead of decrease.

TYPES OF SOURCES

Particles	**Chemicals**	**Living Organisms**
House Dust	Building Materials	Mold
Pet Dander	Cleaning Supplies	Bacteria
Pollen	Personal Care Products	Yeast
Cockroach	Pesticides	Cockroach
Dust Mites	Mold	Dust Mites
Mold		

TYPES OF SOURCE REMOVAL

Removal
Best solution because it is no longer present.
You aren't being exposed.

Isolation
If it can't be removed it can often be blocked.
It's still present but you aren't being exposed

Dilution
Ventilation. It's still there but at reduced levels.

Reduction
Filtration. It is still there but at reduced levels.

SOURCE CATEGORIES

The starting point for sources is to categorize them. You can consider such known sources as cat dander, dog dander, mold, house dust, pollen from trees, grasses and weeds, chemicals, fumes, fragrances, odors, cockroaches and dust mites. (Food allergies and intolerance can also cause complaints. Although those reactions are not a primary focus of this book, it is important to be aware of them.)

These traditional categories help but are still too complex. To further simplify sources, we need to establish a new starting point. Instead of trying to identify every possible source, we can categorize them by basic physical characteristics. Those same characteristics also conveniently determine how to effectively remove each category. The three categories are:

Particles. Even though most particle sources are too small to be seen, they still have a physical size and can be physically removed. Because removal is based on the size of the particles, it makes no difference whether the particles are dust, pollen, dander or anything else. When you remove one kind of particle you remove all others of the same or larger size. Likewise, the equipment or the technique used determines only which *sizes* of particles will be removed, not which *kind* of particles.

Chemicals and Odors (or Molecules). Molecules have a physical size also, but that size is many thousands of times smaller than the smallest particles. In addition, chemicals can interact with each other, creating new chemicals. Therefore, once molecules of chemicals escape from their container and become airborne, they require very different removal methods than those used for particles. Rather than being based on size, chemicals and odors can be removed with a variety of absorbents or reactants, which are determined by the type of chemical involved. Also, many chemicals are absorbed through the skin, as with cleaning products and personal care products. The only removal technique for this type of exposure is to remove the product from the habitat in order to prevent the exposure to begin with.

Living Organisms. Sources that are alive, like cockroaches, dust mites and mold, can reinfest their location. It is critical to first kill them. Then they can be removed like any other particle. Some living organisms also generate chemicals (VOCs) as they metabolize their food. However, once dead, that chemical source is also gone.

LET'S LOOK AT EACH OF THESE SOURCES IN DETAIL:

Particles

The key to understanding particles is:

1. Particles have a specific source; they don't just magically appear.
2. Particles accumulate in some surprising places and can then become airborne again.
3. Particles, including allergens and respiratory irritants, have a physical size. However, about 98 percent of them are microscopic and are too small to see with the unaided eye.

4. Particles can be physically removed from the air. Which ones are captured and which are not depends on the filtration materials and techniques.

5. Exposure to particles typically occurs through direct contact and by respiration.

Reducing exposure is accomplished by removing the indoor sources of the particles and by keeping outside sources outdoors. Successful removal is largely determined by particle size. As the illustration shows, air is forced through a strainer, or filter media. The media has holes in it that allow the air to pass through. However, anything that is smaller than the hole will also pass through. So, based on this information, filter media and vacuum bags should have holes, or pores, small enough to stop the smallest offending particle, typically anything larger than 0.5 micron.

FILTER STRAINER

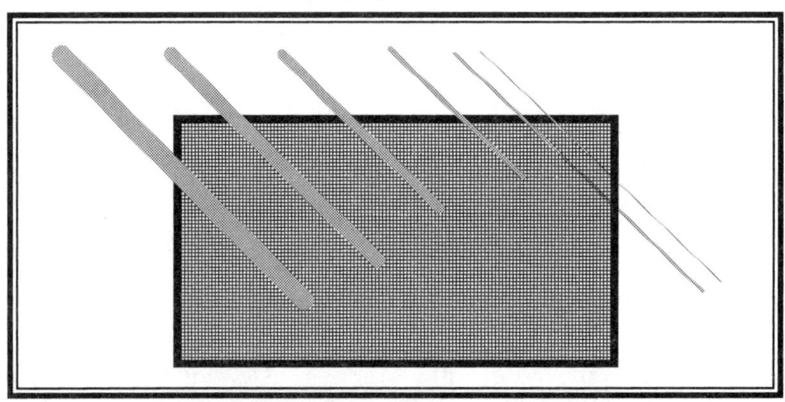

If the particle is bigger than the pore size, the hole through the vacuum bag, then it will be stopped. However, if the particle is about the same size, some will go through and some won't. Particles smaller than the pore size go straight through, as if no barrier was present. If you want to stop common allergens and irritants larger than 1.0 micron, the pore size must be 1.0 micron or smaller.

SPECIFIC PARTICLE SOURCES

Let's now apply what we know about particles in general to the specific allergens and irritants we want to remove.

Cat dander, obviously, comes from cats. Remove the cat and you have removed the source. However, you have *not* removed the *accumulation* of cat dander. If you don't remove the accumulated dander, your exposure can continue for years. Because the particle size of cat dander is very small, it easily becomes airborne and can travel everywhere in the house, distributed by normal air movement. The dander will accumulate on walls and ceilings and on the tops of high objects, such as picture frames and draperies. If you have a forced-air system, then the dander will also accumulate inside the ducts. It can also accumulate on your clothes, your bedding, and your body. Air purifiers with small pore sizes are quite effective at removing cat dander, but only from the air that *actually* goes through the filter. It won't remove the dander from anywhere else. Vacuums can be effective, as long as they have a proper bag. However, cat dander is somewhat "sticky" and is difficult to remove from carpets and upholstery, even with repeated vacuuming. Even carpet shampooing will not remove all the dander. Because of the persistence of cat dander, it is likely to linger long after a home is sold, becoming a potential problem for the new owners. One particularly confusing characteristic of cat dander is that not all cat dander is alike. Two primary types of dander have been identified, which means that some allergic people may not react to all cats.

Dog dander, of course, comes from dogs. Because dog dander tends to be larger and heavier than cat dander, it is slightly easier to control. It will still travel throughout the house with normal air movement, but tends to stay more localized and to accumulate on lower surfaces. Dog dander will be on walls but less so on ceilings. It will also accumulate on your clothes, your bedding and your body. The same statements about purifiers, vacuums and cat dander apply to dog dander. Also, dog dander seems to degrade more quickly than cat dander. And as with cat dander, not all dog dander is alike. In fact, *five* different types of dog dander have been identified, which can further confuse attempts to specify dog dander as a source of exposure.

Pollens are seasonal, occurring only when the host plant is reproducing. Large quantities usually occur outside the house, rather than from indoor plants. Pollen enters a building through the windows, on pets, and on your clothes and hair. When pollination is active, your exposures indoors are near open windows and from your pillow. Pollen attaches to your hair, you lay your head on the pillow, and pollen falls from your hair onto the pillow. When you roll over, you put your face right into the pollen. Two effective ways of avoiding pollen exposure is to keep the windows closed and to wash your hair before retiring. Removing pollen from indoors doesn't seem as critical or as difficult as removing pet dander. Pollen tends to degenerate rather quickly, and its large size and weight results in rapid settling from the "breathing zone" of the air onto carpets and furniture. Because of the large size of pollen, vacuum cleaners can remove it without blowing it back into the air.

Dust mites are particles. And once they are killed, they can be removed like any other particle. However, because their primary characteristic is that of living organisms, they will be discussed in detail in the *Living Organisms* section.

Cockroach allergen comes from body parts and their feces, which are particles. However, because their primary characteristic is that of living organisms, the main discussion will be presented in that section below.

Dust is a very broad category. All the previous specific particles are components of "dust." However, this discussion will include other types of "dust." Dust comes from outside, from clothing as lint, and from building materials, especially during construction or remodeling. Sawdust and drywall dust are two of the most common forms. No matter how you try to control the movement of the dust throughout the house, it will still go wherever the air goes. It will accumulate in the forced-air ducts, baseboard heaters, along ledges and door frames, and behind furniture. Slight changes in air movement will disturb the dust and recirculate it. Another source of dust could be a woodworking shop inside the building. Even if it is located in the basement or in an attached garage, the smaller dust particles will still distribute everywhere throughout the house.

OZONE - THE CONTROVERSY

Ozone is extremely effective at killing airborne mold and other microorganisms. It can also help to control chemicals and odors. However, it does have its limits. Not all chemicals can be oxidized. Some only combine with nitrogen.

Also, the use of ozone in occupied areas is very controversial. The manufacturers and salespeople are on one side of the argument and public health and safety officials are on the other. Some manufacturers and the federal government have countersued each other. Individuals tend to argue one extreme or the other, often with increasing vigor and invective.

Argument: *Ozone is only a molecule of two oxygen atoms (O_2), with an extra atom attached (O_3). How can oxygen be dangerous? The only change would have to be for the better because more oxygen is now present. In fact, people with asthma and other respiratory ailments can only benefit from this supercharged oxygen.*

Counter argument: *Just because ozone is made up of oxygen doesn't mean it continues to act the same as oxygen. In fact, ozone is an unstable, chemically active molecule called a free radical. Free radicals attempt to combine with whatever comes into contact with them often damaging the body.*

Counter-counter argument: *A balance of free radicals in the human body is essential to health. Some critical nutritional processes require free radicals.*

Counter-counter-counter argument: *Carbon, also, is necessary for life. Carbon is the basis of our life form. So why not add carbon to water and get super water? Water is H_2O and adding carbon would create H_2OC. But the actual form it takes is HCHO — better known as formaldehyde.*

This debate offers nothing but confusion and hard feelings.

The answer is that individuals do themselves a disservice if they ignore the knowledge of public health. Then they must take personal responsibility to obtain other information as it relates to themselves. You are the ultimate authority on matters of individual safety — and considering the contentious nature of the ozone controversy, this is especially true.

LIVING ORGANISMS

The keys to understanding living organisms are:

1. Living organisms are alive. Therefore they breed, often multiplying their population to extremely high levels. If you don't remove them *all*, they could easily reinfest your house.

2. Once living organisms are killed, they are then just like any other particle and can be removed in the same way as other particles.

3. Some living allergen and irritant organisms, such as cockroaches, are large enough to be visible. Others, like dust mites, are too small to be seen. Mold, as commonly understood, is visible. However, that is true only when environmental conditions are conducive to reproducing rapidly enough to form "clumps" of mold called colonies. When insurance inspectors and damage restoration workers say no mold exists, what they really mean is that they see no visible mold colonies. Individual organisms can occur at quite high levels and still be invisible. For example, during the summer and fall seasons, the pollen and mold counts in newspapers and on the TV news frequently state that mold levels are extremely high. But you don't see clumps of mold floating in the air. The airborne mold is *dispersed* rather than colonized. And airborne mold is much more common indoors than is colonized mold.

4. Although the primary category of mold and other microorganisms is that of living organisms, it is important to know that they can be a source of chemicals called Volatile Organic Compounds (VOCs) that can also cause exposure problems.

SPECIFIC LIVING ORGANISM SOURCES

Dust mites. Dust mite feces are what triggers reactions, not the mites themselves. And because the feces are relatively heavy, they rarely become airborne. For example, studies have shown that this allergen usually doesn't travel much more than two feet from a forced-air vent. Because their favorite meal is the dead skin your body sheds, dust mites will be found where human skin accumulates — where you spend most of your time, like in upholstered furniture and bedding. Dust mites also prefer humid climates. They are rarely found in the high plains of the Rocky Mountains or in deserts. While not proven, there is some evidence that mites will eat only the skin that has first had the fat removed by mold. Kill the mold and the food chain is broken. The mites will starve. Once killed, the mites can be removed like any other particle. Vacuuming, with a proper bag as described in Chapter 10, is very effective at removing the allergenic feces. Also, tannic acid preparations are effective at neutralizing the allergens.

Cockroaches like to eat food we drop on the floor. Standard kitchen hygiene is the most effective method of preventing their infestation. However, they will also eat glue, old adhesive under tile and linoleum, and just about anything else that has organic matter in it. And like the dust mites, once they are killed, they can be removed like any other particle. Vacuuming is effective per the above specifications. If you feel you must spray or fog for roaches, first educate yourself about chemical sources and potential reactions so you don't replace a particle source of exposure with a chemical one.

Mold is a more difficult problem. The types of mold that are most often responsible for exposures are the airborne molds. These organisms reproduce by filling a little sack (called an ascus) with spores — much like a puffball in the woods. As the sack becomes very full, it begins to bulge and then suddenly ruptures, spewing millions of new spores into the air. Because they are very small and very light, the spores travel on air currents throughout the whole house. They then accumulate on floors, walls, ceilings, drapes, furnace ducts and bedding. If they find a habitat to their liking, they can reproduce, forming a secondary source of mold. They can stay permanently suspended in the air, much like cat dander. The specific type of mold, while important to your allergist, usually isn't as important as the total amount of the exposure to all molds present in your home.

MAJOR SOURCES AND AMPLIFIERS OF MOLD INCLUDE:

- Forced-air ducts, especially if a humidifier is on the furnace.
- Cooling coils and drip pans in central air conditioners.
- Evaporative coolers.
- Carpets and carpet pads that have been wet several times or for longer than two days.
- Building materials that have been water damaged from roof leaks, plumbing leaks, sewer backups and groundwater flooding.
- Damp basements.
- Crawl spaces.

- Food. Even slight microbial growth, at much lower levels than necessary for food poisoning, may cause difficulties for people with acute mold reactivity. This can be easily confused with other sources of exposure. (Again, although food is not a part of this book, it is still important to be aware of this possible source of complaint.)

Ozone - What To Do

Ozone can be quite effective at killing mold and other airborne microorganisms. Used correctly, it can often accelerate the aging of a number of chemical sources, such as paint, sealants and carpets. However, ozone will not penetrate deeply into materials and is therefore most effective with substances that are airborne or on the surface of objects.

I recommend the use of ozone to kill airborne mold, but only after all active sources have been first removed. However, just like everything else we have discussed, this, too, is fraught with dangers and caution is needed.

It is important to find a professional with ozone-generating equipment who will follow these specifications:

1. *Ozone should be generated to levels of at least 10 parts-per-million (PPM) for a minimum of 48 hours, and usually no longer than 72 hours.*
2. *People and air-breathing pets must be out of the house during the entire treatment.*
3. *At the end of the treatment time, the entire house must be ventilated for a minimum of two air exchanges.*
4. *Reentry is allowed only after an additional 12 hours. If the occupants are particularly susceptible, reentry should be delayed for an added 24-36 hours.*

Surprisingly, wet/dry cycles often will generate much higher levels of mold than constant wet conditions. If the environment for the mold organism is constant, the population will increase to a saturation point and then plateau. It may even drop a little bit. But if the environment dries out, the mold strives to protect itself by reproducing, generating millions of new spores. If moisture returns, new spores germinate into new living organisms. If the cycle repeats several times, the levels of mold can become extremely high.

Live mold also generates chemicals called Volatile Organic Compounds (VOCs). VOCs are the type of chemicals that "out gas" from new carpet, furniture, paint, cleaning products, etc., and are most often implicated as causes for Sick Building Syndrome (SBS). However, killing the organisms halts the generation of the chemicals.

Airborne mold can persist as long as eight years after its original source has been removed. This means that once the water sources have been stopped and all damaged materials have been removed and replaced, you can still be exposed to the mold that remains in the air. This is very similar to the presence of cat dander after the cat has been removed. Unlike cat dander, however, mold is alive and can continue to reproduce. It is more like algae in a swimming pool that persists despite repeated eradication efforts.

Mold can also adapt to changing environments. For example, as the active sources of mold are removed, the airborne organisms often adapt to the new environment. Mold can also adapt to more powerful actions. One client, for example, successfully used copper sulfate in his crawl space for about 10 years, only to have the mold eventually reassert itself. Another client tried to control mold in his crawl space by installing bright lights. That worked for a year or so until the organisms adapted to the light.

To stop your exposure to mold, original and secondary sources must be removed. Then the airborne organisms must be killed.

Finally, you may have to remove the dead organisms. Most people stop reacting once the airborne mold is killed, but about 2 percent will continue to react. A common misconception is that dead mold is not an allergen. However, when your allergist tests you for mold allergies, he doesn't use live mold because it would both reproduce and die, always changing the concentration. Instead, the allergist uses mold that has been killed, preserved, and then diluted to specific concentrations. Your reactions are to dead mold.

If you need to kill airborne mold, do not "fog" or spray your home with fungicide, bactericide or disinfectant, especially if you are very susceptible to odors and fumes. Almost all fungicides, including those registered with the EPA, are neurotoxins. Some damage-restoration companies use disinfectants that are mostly formaldehyde, phenol and even gluteraldehyde. These can be very noxious and persistent. Be extremely cautious, even if you aren't chemically sensitive.

Ozone -What Not To Do

Do not deliberately expose yourself to ozone — at any level. Public health information treats ozone as a pollutant. It is the primary reason for the escalation of hospital emergency room visits for respiratory problems when pollution levels are high. Ozone is an oxidant, which means it is unstable, very active, and attempts to react chemically with anything it comes into contact with. That is what makes it so effective in killing microorganisms such as mold. According to most authoritative sources, prolonged human exposure to even low levels of ozone may cause respiratory damage by damaging the cell walls of the respiratory system.

I cannot recommend air purifiers, filters, or any other device that generates ozone in the presence of people and pets. The ozone generated is no different from any other ozone and has the same potential for damage.

If the ozone level is successfully controlled to stay below federally regulated outdoor concentrations, then it is not strong enough to be useful except with continuous use. If those ambient levels were strong enough to be effective, then we would already have a sterile world outdoors. On the other hand, if the level of ozone from the purifiers were high enough to be effective, then the resulting concentration probably would be high enough to violate federal standards and to increase the potential damage to people.

If you are extremely susceptible, especially to chemicals and odors, then I suggest extra caution concerning ozone. In fact, you may not want to consider even an electronic filter for your furnace because they often generate some ozone. It's at extremely low levels, but if you are extremely susceptible it may be high enough for you to react. Some of my clients react to the ozone generated by the electric motors in their refrigerators.

Again, individuals do themselves a disservice if they ignore the knowledge of public health. They must take personal responsibility to obtain additional information as it relates to themselves. Then they must make their own decision about their individual situation.

CHEMICALS AND ODORS

The keys to understanding chemicals and odors are:

1. The size of chemical molecules is thousands of times smaller than particles. For filtration purposes, they essentially don't have a physical size. They pass straight through regular air purifiers as if they weren't there.

2. Levels of airborne chemicals can be reduced by ventilation or by filtration with charcoal, potassium permanganate or zeolite.

3. Filtration and ventilation of an active chemical source does not remove the source. It only reduces the source. For example, if someone is smoking a cigar in the middle of the room, the odor and cigar combustion molecules will still be present between the cigar and the closest window. Even if Hurricane Hugo blows through the room, they still exist, although the levels will be dramatically reduced because they don't have time to accumulate.

4. Stopping exposure to airborne chemicals and odors often is easier than stopping exposure to particles or living organisms. Most chemicals easily evaporate and are quickly diluted in the air. Ventilation is very effective. Open a window and in a few minutes most of the odor is gone. However, highly susceptible people may still react to the occasional molecule they may encounter by chance.

5. When stopping the exposure isn't easily accomplished, chemicals and odors can be difficult or even impossible to remove or isolate.

6. It is estimated that more than 600,000 man-made chemicals exist, with over 60,000 of them in active use. Regulatory compliance is concerned with only about 400 of them.

7. Exposure to chemicals can occur by breathing airborne molecules or by direct skin contact. Unlike particles and living organisms, chemical molecules can penetrate the skin and enter directly into the body. They can also be ingested with food and water.

8. Reactions to chemical exposures are usually not allergic reactions.

9. A chemical does not have to be a toxin for you to have an adverse reaction to it.

Molecules

Molecules are messengers between the outside world and our bodies. But these messengers are not something separate or different, like is implied when we talk about the "odor" or "smell" of something. Odors are actual "pieces" of that object that have broken off; much like what we taste is actual pieces of food.

When you smell a bad odor or one that makes you sick, it's not much different than eating it. When you smell ammonia, formaldehyde or even body odor, "pieces" of ammonia, formaldehyde or that persons body are actually entering your body.

So be as careful of what you smell as you are of what you taste.

Robert W. Rinehart, Sr., Ph.D.
Rinehart Laboratories, Inc.

10. The chemical does not have to violate regulatory law for you to have an adverse reaction.

11. Long-term exposure to low levels of chemicals often is more damaging to individuals than a single toxic exposure.

12. Chemicals are the most common cause of replacing one problem with another. Killing mold with a fungicide or disinfectant frequently replaces a source of living organisms with a chemical neurotoxin. Putting a sealant on water-damaged building materials replaces a microbial or odor source with a chemical one. Simple house-cleaning tasks of removing dust, dirt and grease usually involve the use of several chemical products. Carpet cleaning can be an incredible source of chemical exposure. Even "safe" products usually contain a chemical fragrance.

SPECIFIC CHEMICAL AND ODOR SOURCES

Because of the overwhelming number and types of chemicals and their varied locations, we need to change our perspective slightly. Our new starting paint will be based on where the chemicals come from, using the following categories:

- Building materials
- Cleaning and disinfecting products
- Personal care products
- Food
- Fragrances

Building materials are a major source of chemical exposure. Fiberglass insulation often contains a phenol-formaldehyde resin that holds it in the form of a batt. Particle board and chipboard need glue to hold them together; the resin is often a urea-formaldehyde substance. Some carpet pads contain pieces of car tires. Carpet and floor tile adhesive can take years to stop emitting odors and fumes. Even though latex paint is better than oil-based paint — which is better than the old lead-based paint — it still often contains high levels of Volatile Organic Compounds (VOCs). New carpet and furniture may "off gas" a veritable witches' brew of chemicals.

Cleaning and disinfecting products are, by definition, chemicals. Virtually none of the pesticides and disinfectants registered for use by the EPA have been tested for neurotoxicity, although that is the means by which most of them kill.

When pesticides are added to other substances, such as paint, they no longer are regulated as a pesticide. Banned pesticides such as DDT can still enter the country on fresh produce from foreign countries. DDT can even be used in this country if it is included as an inert ingredient in other pesticides. Even seemingly innocuous deodorizers for the home may contain low levels of pesticides.

Personal care products, especially soaps and lotions, can be a significant source of chemical exposure because they can be absorbed through the skin. Just read the labels on a few of the products. You will find an incredible number of ingredients that can hardly be pronounced, let alone understood as to their effect. Despite all else, the fragrance is often the culprit.

Common products used to remove contaminants, such as vacuum cleaners and cleaning chemicals, are themselves often a source of exposure. But because they are "common" or "normal," they are often overlooked.

Foods can contain chemical preservatives to delay spoilage. Meat and poultry often is "washed" with chlorine bleach or even with a formaldehyde or phenol-type disinfectant. This is critical to help prevent food poisoning and deadly salmonella. However, if you are susceptible to these substances, then you need to know if they are present in the food you buy. This can be difficult because if the level of a chemical is below regulatory requirements for any product or food, it doesn't have to be listed on the label.

Mary was an active, middle-aged, professional woman. But her chronic sinus infections were interfering with her enjoyment of life. Later, she began to miss work because of their severity. The worse the infections became, the more she cleaned — often vacuuming several times a day in an attempt to remove the dust faster than it could accumulate.

An inspection of the house revealed no obvious source of "dust" other than the vacuum cleaner with a standard bag. When Mary stopped using the vacuum cleaner, her sinus infections decreased dramatically and were more easily controlled by medical care.

The dust from the vacuum cleaner wasn't the primary cause of the chronic sinusitis, but it was surely a large influence on the sequence of events that led to the infections.

Fragrances aren't limited to personal care and cleaning products. They can be anywhere. You can buy oven hot pads with little plastic beads inside that are imbedded with fragrance. Vacuum cleaners, whose bags are so porous that you can smell the dust that escapes, often have a little slot to insert a fragrance packet. Even some air purifiers, whose fundamental purpose is to remove pollutants, come with a fragrance! Some shopping centers, especially in Japan, are experimenting with introducing specific fragrances into the air to stimulate buying. Research suggests that a similar technique can increase productivity in business offices. Again, these claims are validated with group studies withing a closed system. Individual complaints are excluded as unscientific anomalies.

Jan was allergic to mold. Her symptoms increased dramatically when she moved into a house with a damp crawl space. However, her symptoms were only partially alleviated after the crawl space was dried out and isolated from the rest of the house. They were particularly noticeable in the master bedroom near the bathroom.

One day she was cleaning the bathroom and decided, as an experiment, to move all the cleaning supplies, soaps and perfumes to another room. As she was carrying the box full of bottles to the garage, she started developing symptoms similar to her mold allergy. When she returned to the master bedroom she was symptom-free. She had found the source of the rest of her "mold" symptoms.

The only effective way to avoid chemicals indoors is to not bring the chemicals in. When that can't be avoided, you are left with techniques that only dilute by ventilation or reduce by filtration. And this can be a very difficult problem.

Chapter Nine

Medieval Monsters Modern Dilemmas

MEDIEVAL MONSTERS - MODERN DILEMMAS

This fearsome, medieval monster is a dramatic symbol for the particular type of contradictions and dilemmas you will face on your quest for a complaint-free indoor environment. It is so ugly and repulsive that you would much prefer to ignore or deny it's presence than to face it.

This monster is not only fearsome looking, it has a unique ability to confound any attempts to defeat it. Because of its many heads, the natural tendency for any warrior is to cut them off one by one. But as each head is removed, two more immediately replace it. The harder you fight the obvious enemy the worse your situation becomes!

A new tactic is needed to defeat these frustrating creatures. But in the meantime, it will be helpful to identify and demythologize some of these modern-day monsters so you can recognize them for what they truly are: a rallying point for new tools and new methods.

Dilemma #1

What do I do while waiting for science to provide definitive answers?

Are pesticides and disinfectants dangerous to me and my children? Damage restoration companies insist that if the fungicide is EPA registered, that means it is safe. Pesticide application companies claim that their products are safe. And then I see pictures of all those deformed alligators in the swamps of the south and of the frogs up north. Could it be that nobody really knows? How many times do I need to wash the fruit before all the pesticides are gone?

One doctor says I have chronic fatigue syndrome. Ten others say it's a phantom illness. Yet they can't agree on what is wrong with me. However, the only doctor who is giving me any relief at all is the one who is treating me for chronic fatigue syndrome. Does that mean I have it? Is it just a coincidence? Or am I just imagining the whole thing?

How many respiratory attacks does my 12-month-old son have to have before he has asthma? Some doctors say three, some say five. But all I want to know is what do I do? I don't care if technically it's asthma or if it's caused by an invasion of alien creatures. What can I do to help him? Now!

Dilemma #2

How do I make effective decisions when the available information is incomplete or contradictory?

One day I hear that red wine prevents cancer and the next day a new study says that it doesn't. Which is worse, cancer in 20 years or alcoholism today?

One study says I should eat broccoli and another disagrees. I don't like it anyway, so I won't. So what if I die one day early.

One day the earth is in danger of getting too warm and the next day it isn't. One day I should reduce fat by avoiding butter and eating margarine; the next day a new study says that the fat in margarine is even worse. But don't go back to butter!

I know smoking is bad for me and that secondhand smoke could really harm my kids. But if I don't smoke I'm grouchy and irritable and I spank my children more often.

Dilemma #3

How do I know if my home is safe if there are no definitive tests? And if there aren't any, then how do I know what to clean up — and if it has been successful?

I thought I could smell a gas leak from the furnace. The utility company tested and said it was safe. An industrial hygienist said there was "none detected." I called my heating contractor. He said there was a leak but not enough to cause an explosion. On my insistence, he replaced the slightly leaky valve. I don't smell gas anymore. Out of curiosity, I had the utility company retest. They said it was safe. The industrial hygienist said there was "none detected." My (lack of) migraines said it was "gone."

After we moved into a new house, my asthma got real bad. A pre-purchase inspection had found "nothing wrong." Even the radon was very low. Then a specialist tested for dust mites. We cleaned up the house and the retest showed they were gone. But my asthma hasn't improved. Cat dander can trigger my asthma also. But we know of no test for cat dander and the previous owner insists they never had any cats during the six months they lived here.

Dilemma #4

How do I cope when no one else reacts the same as I do?

I've lost a lot of friends lately. We used to go country western dancing every weekend and it was my favorite activity. But recently the cigarette smoke just wipes me out. I get tired after an hour and have trouble breathing. All my friends are having a good time but I'm miserable. I haven't gone dancing with them for a month now and they haven't called me for a week. I miss them.

Dilemma #5

How do I heal if no one else believes me?

I've been working hard on avoiding exposing myself to anything that causes problems — such things as dust, mold, perfume, and dogs. When successful, I feel so much better. But the man in the cubicle next to mine at work wears such strong after-shave lotion that I can smell him from halfway across the room. I've asked him to use less but he just looks at me like I'm from Mars. I need him to believe me so he'll cooperate.

Support systems are wonderful. They enhance my efforts at achieving new goals and exploring new parts of myself. But my support group doesn't believe that I truly get sick from their (chemically) clean houses and from the car exhaust on the highways. When I try to explore that part of me with them, they gently tease me about how I need to be more positive and not let such insignificant trifles get me down. I want to be with them but I no longer feel a part of the group. Oh how I'd love to have their support! Instead I'm becoming angry and spiteful. And I don't think that is very healthy.

Dilemma #6

How do I heal if the people around me are hostile?

Relaxation and meditation are very helpful when I am in reaction. But every time I start, my husband suddenly has to get something from the room, and he makes comments like, "Oh, are you sick again? What is it this time?" Sometimes at work, people play tricks on me. They bring me a pile of papers to work on but first they spray them with perfume. I even get attacked at my doctor's office. On the last visit I had been telling the staff about how clothes that had been dried with fabric-softener dryer sheets give me a migraine. So this visit one of the other doctors is walking around with a fabric softner sheet hanging out of his pocket, asking others if I've started complaining yet!

Dilemma #7

How do I heal if I cannot find the balance between denial and hysteria?

If I ignore my symptoms, I just get sicker. But if I pay attention and avoid being exposed I get much better. But then I'm accused of over-reacting and am labeled an hysterical hypochondriac.

Co-workers and friends have begun showing me books and newspaper articles which claim that people with my condition are no different from those who claim to have been abducted by cults or aliens. That is so outrageous that I do become hysterical!

A Common Dilemma

So your house is making you sick. You are tired of all the sinus infections, allergies, sneezing, the looming threat of an asthma attack, fatigue or headaches. You've decided that your starting point is to remove the source of the problem by cleaning your house.

It sounds simple enough. Just avoid the irritants, triggers, and allergens by getting rid of your beloved family pet or by becoming a better housekeeper or paying a premium for a special allergen cleaning crew or spending $3,000 for comprehensive laboratory testing. You can also research the Environmental Protection Agency regulations, call OSHA and your local health department, or hire a Certified Industrial Hygienist. At the other extreme, just clean your forced air ducts and your carpets. And buy the latest HEPA air purifier. Or are ionizers better? Or is it ozone?

While you are at it, get a better vacuum cleaner. Oh, yes, don't forget the barrier cloth for your mattress and replace that old musty carpet with hardwood flooring. But wait... the polyurethane floor finish makes you and your child feel sick. And if you do nothing else, replace your forced-air heating system with baseboard heat. Ignore the expense. Your health is worth it! Maybe it would be easier if you just moved.

But don't all houses have problems? What if the new house is worse than what you already have? Maybe you should just stay where you are. After all, allergen exposure isn't life-threatening. Just learn to live with it.

Because the exposures and discomfort haven't stopped, you take further action. You start reading several of the numerous excellent books available. But as you read you become overwhelmed with information. Also, some suggestions work but others don't. Several are very costly, perhaps beyond your budget. And do you have to do *all* those things? Which ones can you skip?

You may have successfully stopped your exposures, only to have them recur. Or you may have given up long before this point, unless your discomfort is so great that you become even more desperate to find relief.

What should you do? Whom should you ask? Whom do you rely on? How will you know if your actions are successful? Who does the testing? Where's the instruction manual? You need more information before you can even determine a starting point that will allow you to succeed.

WHAT DOESN'T WORK

Let's start with some simple observations about what *hasn't* worked. You would *not* have indoor exposure problems if all that's necessary is:

- Ordinary housecleaning
- Laboratory testing and industrial hygiene procedures
- Using air purifiers and better vacuum cleaners
- Complying with local and national public health policy

All of these actions may help. But, at best, they solve only the simplest exposure situations. If this has been your experience, then your situation is more complex. Although the techniques for solving complex problems are relatively simple, the "how" and "when" to use them can quickly become dumbfounding.

To comprehend what is actually happening, we need to slightly change our understanding of a few simple assumptions and then modify our starting point. Imagine a situation that is not life-threatening but is still rather disturbing. For example, you feel something touching your leg under the table. You don't know what it is. It could be your spouse caressing your leg. It could be your cat. It could be a bug. Or a bee. You can't identify it so you don't know what to do. Because you need more information, you ask your spouse if she is touching your leg. If she says she is, you know what is happening to you and you know what to do.

But if she is not touching you, then what is? Will it hurt you? What if it's a cat that can scratch or bite you? Or a bee? Or a snake? You gather more information by literally shifting your position to look under the table. If you still can't see the source of the touching then you need to shift your perspective again. You keep doing that until you can "see" what is happening. If it is a bee, you know what to do and you are not unduly afraid. You can relax. But if it is a snake, you might panic. You may find the *idea* of a snake so repulsive that you don't want to even think about, let alone dare to look to see if a snake is actually present. Despite your fears, your need to know whether the snake is there or not is critical for determining what action to take and what action to avoid. Being in denial about a possible encounter with a snake leave you vulnerable to ongoing harm.

What you *don't* need is a laboratory instrument (even if one were available to detect objects under tables) to identify the problem. You don't need the most recent results on a double-blind, controlled, scientific study. You don't need statistical data about what is under the table in other households. While that information may be helpful, you don't necessarily need it in order to successfully identify what is touching you under the table and then to decide what you want to do about it. Right now. Allergens, respiratory irritants, headache triggers, brain "foggers," and asthma-causing substances can be as subtle and as mysterious as "something" being under the table.

However, the vast majority of indoor allergen, irritant, and chemical exposures can be dramatically reduced, if not stopped altogether, by simple procedures. But *not* if you begin your journey from traditional starting points.

HERE ARE FIFTEEN REASONS WHY:

1. First and foremost, almost all indoor exposure situations have either already been fixed or prevented. That is why the vast majority of the population has no complaint. But the fact that you are still reading suggests that this is not the case for you. *Therefore, your situation is an exception to what is normal and it requires additional information and new ways of perceiving what is happening.*

2. You don't always need to know the precise allergen, irritant or other substance that is affecting you. Most of them, for purposes of both identification and for removal, can be placed into the three general categories of particles, chemicals, and living organisms.

3. Successful removal techniques are relatively simple. "Rocket science" is seldom necessary. But knowing where and how to use those removal techniques and knowing what to avoid can be complex.

4. The complexity of non-standard situations comes from the combinations of multiple sources, delayed reactions, strong reactions to low levels, and the individuality of your experience.

5. There are no standards yet, regulatory or otherwise, for common allergens and respiratory irritants. Less common substances and nontoxic chemicals are considered inconsequential. Therefore, regulatory agencies are usually of little or no assistance.

6. Allergic reactions and other similar types of individual, nontoxic reactions and irritations are not considered a public health threat. They do not cause cancer and are not contagious like AIDS or physically harmful like a car crash. Therefore, public health departments and public health regulations can usually offer only limited assistance for individual complaints.

7. Not all reactions are allergic reactions. Nonallergic reactivity is fairly common. ***Your*** reaction may not be an allergic reaction.

8. The combination of all your reactions is unique to you. No one else has quite the same experience that you do.

9. Statistics describe a group. They say absolutely nothing about a specific individual. Just because the statistical evidence shows no correlation between an exposure to a substance and your reaction, that does not mean that ***you*** are "safe." Therefore, your unique experience is not necessarily a hallucination or other form of mental illness.

10. Most low-cost — and even some very expensive — vacuum cleaners, air purifiers, and furnace filters are designed to remove visible dirt. However, it is the much smaller, **subvisible** "dirt" that is of primary concern. In fact, standard vacuum cleaners are one of the greatest indoor polluters.

11. Most common cleaning products are excellent for removing dirt, grease, and grime. However, they also contain fragrances, chemicals, and occasionally pesticides. They can easily replace one source of exposure with a chemical one.

12. Product manufacturers and cleaning service companies provide the majority of the information about what removes pollutants from indoors. That information is fairly accurate as to how it accomplishes its advertised function. However, any claims of safety are usually based on a self-interest in selling their product or service, within the scope of current labeling laws and on statistics that apply to the general public. They say ***nothing*** about individual safety — ***your*** safety.

13. No one knows what is "safe" for you. So don't ask, "Is it safe?" Those you ask don't know. They can't know. Your family and your friends can only tell you their experience. And *their* experience is that it is safe. If they didn't truly believe that, they would be inundating *you* with their horror stories!

14. Salespeople don't know what is "safe" for you. So don't ask a salesperson if a product or service is "safe." How would you expect them to respond? Would they tell you, "No, don't buy what I'm selling, because it's dangerous"? Of course not. If they really believed that it was dangerous, they would either quit their job or be lying to you.

15. What about your doctor — especially if he is an allergist or a specialist in environmental or occupational medicine? They are a critical source of health information. He or she can advise, and often diagnose, many of the substances to which you are medically susceptible. However, physicians cannot medically diagnose *everything* that causes reactions in your body. And most importantly, they cannot tell you what you are being exposed to, let alone where or how you are being exposed. How to remove those sources is not a medical matter.

So who's left to consider? We've eliminated just about every possible resource. Whom do you rely on to make these critical starting-point decisions?

The biggest single mistake, even by experienced people, is to ***rely on the expertise or authority of others*** rather than ***making their own choices*** about what to do and what to avoid. Exposures to substances that are not part of our public health and safety domain cannot even be "seen" by those authorities. They cannot see them because they don't yet have the tools. There are no credentialed or licensed authorities in this field. And public issues are not even intended to address individual needs.

The only appropriate course is to rely on ***information*** from others, but not on the *authority* of those who give you the information. If the information comes from family and friends, what is true for them may not be true for you. Gather the information from others, but it is you who must make the decision. And as unfortunate and unjust as it seems, the more desperate your situation the more true this becomes.

The fact is, *you* are the authority. The ultimate authority. Get all the information you can from all the sources you can find. Ask questions, request documentation, insist on references. But *you* have to make the decision about *your* safety. You are the ultimate authority for choosing what to do and what to avoid on an individual, rather than a public, level.

FURTHER COMPLICATIONS

We have a strong desire to just "follow instructions." We say, "Just tell me what to buy and I'll buy it. Just tell me what to do and I'll do it." If your situation is that simple, then you don't need this book. You don't have a problem.

However, if you are struggling, then you know that standard approaches have not worked for you. *This usually means that you are highly reactive to very low levels of exposure, perhaps more so than anyone else around you.*

You also know, firsthand, the incredible frustration, confusion, and even hostility that comes with unsuccessful attempts to stop adverse experiences which other people around you don't share. It may remind you, for example, of when you were a child being pestered by the schoolyard bully. He (or she) pokes you in the back. You ignore it. He pokes you again. You quickly turn around to see who it is but cannot. Then you feel another poke. This time it's hard enough to hurt.

"Who did that?" you cry out. "Did anyone see who did it?"

"Did what?" the others respond. They did not see anything.

You are hit again, hard. Still, no one sees what happened to you. No one else is getting poked. And no one else seems to care. In fact, they start laughing at you and calling you a baby! You protest to your teacher and are told to just ignore whoever is bothering you. "Grow up," you are admonished. "You just encourage them."

But the abuse continues. You grow more desperate. You can't make it stop. Attempts to ignore it just seem to provide more encouragement. People stop believing you. Your friends don't act like friends anymore. They become hostile. They start blaming you for your own problems. Ultimately they reject you.

"Go away," they say, "We don't want anything to do with crazy people. In fact, you are now making *us* crazy!"

As adults, we seldom experience physical events like the example above. But a similar dynamic can occur. Imagine that you perceive something harming you, but no one else can detect it. The subjective and the emotional components of your experience now become critical, often dominating and distorting the very real physical factors. This is normal when there is a scarcity of reliable information about a very personal experience. It raises powerful questions of what is real, what is "crazy," what is safe, who is right and who is wrong. It can disturb your belief system, shatter your sense of inner peace and violate your sense of security.

It is particularly destructive to relationships. One of the "glues" of personal relationships is the commonality of experiences, the familiarity of behavior that often leads to finishing each other's sentences, or even forms of subtle, nonverbal communication where each knows what the other is thinking and feeling.

But what if that commonality were no longer true? What if your spouse, friend, child or parent starts complaining about the very activities you both enjoy? What if they no longer want to go skiing, biking, to the movies or to dine out? What if they now want you to stop using your favorite perfume or after-shave, and then become enraged when you don't? For example, if you are the one with the sensitivities, how do you remove the mold that is making you so ill that you can't even go near it, let alone remove it? And nobody else thinks it important enough to do it for you.

Is it any wonder that many couples who are struggling with chronic, unresolved health problems also have relationship problems? One frequent conflict is a power struggle about who is right and who is crazy. And it affects both people. We often forget that chronic problems of any type have powerful effects on more than just the victim. Children are especially vulnerable.

> *My mom is very sick and it's so painful to watch her suffer. I can hear her in the bedroom weeping and sobbing, as if everyone in the world has died. Somehow, I believe it's all my fault; nothing I do is right. I'm so afraid — afraid she won't get well — afraid I'll be left alone. What if I get her illness too?*
>
> *Diary, I get so angry, but I don't know who to be angry at. I want our lives the way it used to be. I want our family back together again. I wish I had someone to talk to, but the kids at school think we're weird. I've tried talking to God, but I don't think I know how to pray. I'm having a hard time eating and sleeping. I feel like I'm in a dark maze and completely lost. Mom says it's just for now and it will pass. Diary, I don't know. I feel so hopeless.*
>
> *Lari Jacqueline*

Or you may receive a gentle suggestion from your spouse, friend, or doctor that perhaps a therapist would be helpful. While that may be true, what almost always happens is that your *medical* problem becomes perceived as a *psychological* problem. And this can trigger psychological issues.

The tragedy is that all these psychological issues now become officially accepted as the obvious cause of your medical problem. Everyone, including the professionals, usually gets it twisted.

Not only are you the victim, but now you, instead of the "bully," are being blamed.

To avoid being trapped by the multi-headed beast of the Dilemmas, *you* need information that makes sense to *you* and to those involved in *your* life. *You* need information that *you* can rely on, even in the "heat of the moment," so *you* can assess where *you* are in relation to where *you* choose to be. Then *you* need to use that information to ***reestablish and modify*** *your* starting point in the midst of *your* journey. Specifically, *you* need to learn:

- How to determine the most likely cause of your exposures.
- How to determine your Personal Impact Rating (PIR).
- How to develop and execute a personal plan.
  - How to determine the correct sequence of actions.
  - How to remove, isolate, and reduce exposure sources.
  - How to evaluate the results.

- Have you have successfully resolved your complaints?
  - Are your efforts fruitless?
  - Should you move to another house rather than try to fix this one?
  - Is the house even the problem?
- How to avoid the trap of being victimized.

The following chapters develop the information and concepts necessary to make those choices. In Chapter 15 you will find a comprehensive action plan.

Chapter Ten

Evaluating Products and Services

Evaluating Products and Services

If sources and their removal are the markings on a map, then the cleaning products and services are the means of transportation. To travel from London to Boston, for example, you can fly or go by boat. There are a variety of airlines and cruise lines to choose from. You also can build your own craft or go by hot air balloon. Your choices depend on a wide variety of factors, including whether your goal is to travel inexpensively, travel in luxury, or set a new record of some sort. If you aren't clear about your goals and reasons for traveling, you can easily become confused and fall prey to misleading information. And if you don't know your starting point, who knows where you may end up!

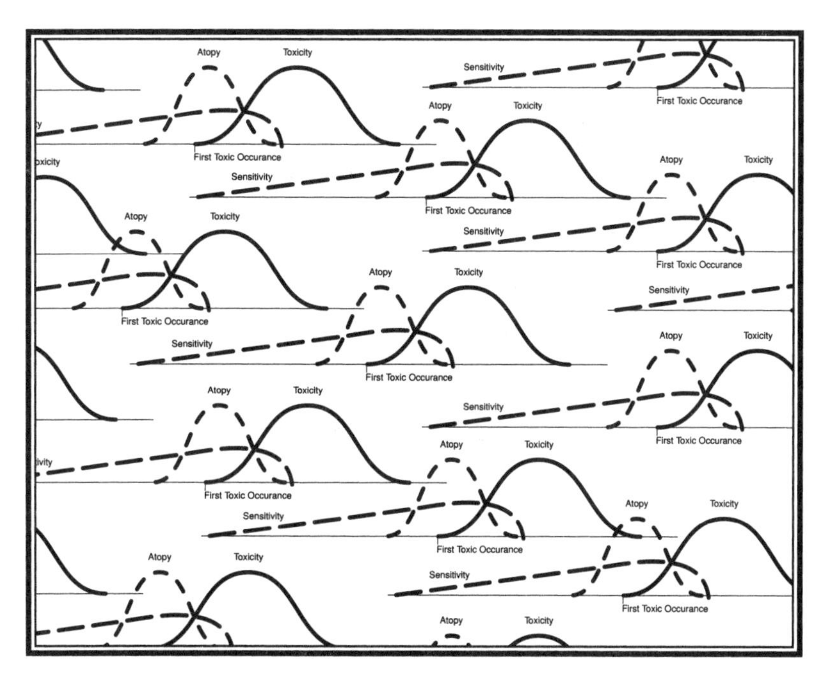

What are your exposure levels to which sources? Where is your susceptibility location on each curve for each source? What is your overall PIR and your PIR for each source? What products and services will actually remove or reduce those exposures to below your reactivity threshold without replacing the original source with a different one?

Evaluating products and services for removing sources of indoor exposure can be very confusing. The methods you choose may not actually remove the targeted substance. They may only be partially successful. They may redeposit it elsewhere. Or your selected product or service may remove the offending substance only to replace it with a different complaint source.

As described in Chapter 8, product and service specifications and benefit claims are typically not much help. Such information often seems aimed at convincing you to give your money to one company rather than another. Furthermore, these companies are not concerned with protecting a specific individual. How can they? Everyone is so different. They'd have to design 250 million versions for a population of 250 million people. That is not our country's basic economic principle of mass production. One size fits all, or you don't get to own it.

AN ALL-TOO-TYPICAL SCENARIO*

Traditional starting points for indoor cleanups are typically very successful. However, when they fail, all further attempts to obtain accurate information about how to remove indoor sources of exposure can be extremely frustrating.

Whom do you ask? If you ask air filter companies, they claim their product is all that is necessary. Ask air-duct cleaning companies and they will say *their* service is best. And then the carpet cleaners claim that without their service you are just wasting your money.

You want help and you need to start somewhere. Lacking direction, you make a guess. Start with carpet cleaning. You ask the carpet cleaning salespeople if they can remove allergens from your carpets. They ask you, "What are you doing with allergens in your carpet?" they quickly add, "if you make an appointment right now, you can get a special offer. Five rooms for only $50" — unless, of course, the carpets actually are dirty, then you'll be charged $300. You discover that there are few industry standards and every company claims to have the exclusive secret of getting carpets clean. Each sounds competent and professional, yet no one understands *your* concerns about being exposed to the fragrance in the cleaning agent. Some

*These hypothetical scenarios are intended to illustrate how a sufferer often perceives his or her experience with physicians and product and service providers. Hopefully both sides can learn from these fictional dramatizations.

even insist on using dry-cleaning chemicals, despite your protest that they are one of your major problems. They also can't promise that the workers won't sneak a cigarette inside your house. For the time being, you give up on cleaning your carpets.

Next, you look up an air-duct cleaning company. These companies usually claim to know what allergens are and agree on how critical it is to remove dust mites from your forced-air system. (Dust mites aren't a problem in all areas of the country and, in fact, they are probably the least important allergen in your ducts.) Each also claims to have the exclusive secret for cleaning. Yet, again, no one understands your concerns about being exposed to the disinfectant they want to use. They insist that it has to be safe because it is registered with the Environmental Protection Agency. When you tell them it makes you feel sick, they look at you incredulously. You also give up on duct cleaning for now.

In desperation, you just buy an air filter. After all, there are industry standards to guide you. But, as your research progresses you discover that there are *six* standards. And they are all different!

Should you get a purifier that meets the standards for the initial dust spot test or the one with the DOP standard? And what is DOP anyway? You have heard that HEPA filters are the best, but they are more expensive. Then you find a cheaper filter that states it is HEPA-like. How much like a HEPA does it have to be to actually be a HEPA? Another manufacturer advertises that its filter is a 95 percent HEPA. You really don't need 99.97 percent, do you? 95 percent ought to be good enough. Shouldn't it? Someone else claims that electronic filters work even better than a HEPA. And you never have to clean it because when it gets dirty it makes a "zapping" sound that removes the dirt. Others tell you that the "zapping" sound means it's long overdue for a cleaning and is now producing harmful ozone. Wait a minute. Didn't you just see a purifier that generated ozone?

But you aren't done yet. You find filters with ionizers, ionizers with ozone generators and superconducting secret-field generators that don't produce either ions or ozone. Their salespeople assure you that they work better than all the others. When you ask which of the six standards they use to test them, you are told that these devices are so advanced that they can't even be tested with any instrument available. They are so advanced that the factory can't even determine whether they are working before they are shipped. They depend on their customers to tell them if there is a problem.

Everybody you ask, it seems, says they are the only ones who can actually save your life. They don't even know your problem (because they don't ask), but they assure you that if you just buy their product, you will be fine. "*Trust* me," each says.

And for those of you who are most severely impacted by your exposures, it gets even worse. You feel exhausted. Your brain is "foggy." You have to force yourself just to dial the phone. Your muscles ache and your head hurts. How are you ever going to sort through such a mess? You can't stand even the *thought* of getting more information!

Out of desperation, you start with an air filter. You are confused about all the claims so you just choose one. When you turn it on it fills the house with a floral fragrance. You find a packet of fragrance inside the cabinet. The odor gives you a migraine, and another false start.

Next you choose a company that cleans both ducts and carpets. Dealing with one company has got to be easier than dealing with two. To avoid repeating the previous mistake, you tell them about your medical conditions and plead with them not to do anything that will harm you. They readily agree. In fact, you have done such a good job of convincing them of the seriousness of your malady that they try extra hard to help you. They use the extra-strong carpet cleaning agent and spray their most powerful disinfectant in the ducts. Your whole house is now contaminated.

And you feel even sicker.

Where do you go now? Your indoor exposures now are worse than before you started. If you can't go home when you are sick, where ***do*** you go?

FIRST THE PROCESS, THEN THE TECHNIQUE

Typical starting points for evaluating products and services are based on techniques and the equipment to be used. However, if that approach fails, it is necessary to reestablish your starting point.

First, keep your focus on your *goal*. You want to stop being exposed to particles, chemicals and living organisms that cause an individual complaint.

You don't really care *how* you stop the exposure. You will use any technique, tool, device, service, or procedure that results in success.

I had mastered being reaction-free in my home even though my husband wasn't very cooperative and my overall PIR varied between 3 and 5. I even hosted a Christmas party for 50 people with little difficulty because the guests honored my request to not wear perfume, cologne, after-shave or clothes washed with scented laundry detergent.

But all that changed when my daughter returned from college wearing a heavy perfume and other products with a fragrance. It made me quite nauseated, agitated and critical of whoever happened to be within striking range.

My daughter promised to not use the perfume but kept "forgetting." Rather than repeat the old pattern of another useless and hurtful confrontation, I packed my bags and left the house.

I checked into a hotel that was used by a local hospital for asthma patients. Unfortunately, the rooms had been recently cleaned with something that had a strong fragrance. I grew even more irritable and uncomfortable.

After several unsuccessful hours and dozens of hotels, I returned home to a tearful and very contrite daughter who promised to never use perfume again. And she didn't — for a few days.

Another discussion took place, outside, so I could be near enough to her to talk. Then my husband came outside and wanted to know what we were doing out in the cold. They both thought I was being too dramatic and extreme.

It took several months and a national television program to convince them that I really did become ill even when my daughter used "only two drops" of her fragranced personal care products. They now understand that my reactions are legitimate and that I am serious when I say I need their help.

The point of this story is that I was willing to leave my family in order to stay safe and healthy. And now I don't have to.

Second, make removal decisions based on the *effectiveness* of those products and services. These are, in order of effectiveness:

- **Remove** the source. It's gone, so the exposure can no longer occur.
- **Isolate** the source. It's still present, but it is blocked from coming into contact with you.
- **Dilute** the source with ventilation. It's still in the air, but at lower levels.
- **Reduce** the source with filtration. It's still in the air, but at lower levels.
- **Combination** of the above techniques. Absolute removal and isolation are rare. If you are extremely sensitive, or if the source continues intruding into your home, you may need ongoing reduction with ventilation or filtration.

Third, base decisions on an evaluation of a *process*, not a specific technique or technology. You don't care that one vacuum cleaner, for example, has airflow of 200 miles per hour at the head, or that another one creates hurricane Hugo inside the collection bag, or that a third can lift a ten-pound bag of hockey pucks. Nor do you care if a duct-cleaning company uses a 4-inch diameter or a 12-inch diameter hose with a shop vac or a huge truck, or if their fungicide is registered with the EPA or with their mother-in-law.

Fourth, in addition to the effectiveness of the product or service, it is equally important to know how it will affect you. What good is the greatest fungicide if it makes you sick? Why install the most luxurious carpet if the odor drives you from the house? Is it important for the bathroom floor tile adhesive to last 20 years instead of 10 if you can't stay there long enough to bathe?

Finally, whose best interest is being served? If the product or service is not customized to meet your individual needs, how can your best interest be served? If it is not your best interest, then whose is it? And why would anyone want to put out the time, effort, and money to solve an intrusive or disabling condition when someone else derives the major benefit?

You aren't interested in the techniques ***themselves*. You are interested in a *process* — one that uses whatever tools, products and**

services are available to help you achieve your goal, which is to stop being exposed by removing, isolating, diluting or reducing sources.

If the service provider can't supply the information you need, then find someone else. How will you evaluate your chances of success if you are working blind? We have enough trial and error in this process the way it is. We need as much reliable information as we can generate.

> *You aren't interested in the techniques **themselves.** You are interested in a **process** — one that uses whatever tools, products and services are available to help you achieve your goal, which is to stop being exposed by removing, isolating, diluting or reducing sources.*

TECHNIQUES FOR THE PROCESS

To simplify cleanup procedures, let's place everything into five technique categories. Then let's discuss what you need to know, how to avoid being misled and how to protect yourself if the workers or the product replace the original offensive source with a different one. With basic information, you can make accurate decisions and will be less persuaded by the others. The categories are:

- Vacuum cleaners
- Air purification devices
- Cleaning products
- Disinfectants and pesticides
- Duct cleaning services
- Carpet cleaning chemicals and services

VACUUM CLEANERS

Vacuum cleaners are usually the biggest single source of particle and living organism exposure inside a home.

Vacuum cleaners originally were designed to remove *visible* particles. Anything smaller than 8-10 microns, which includes most common allergens, not only is too small to see with the unaided eye but can easily blow right through the vacuum bag like dust through a window screen. The assumption was — and often still is — if you can't see the dust, then it isn't there.

Now we know better. We also now know that we need a vacuum that will not only clean the carpet but will *not* allow the subvisible allergens and irritants to blow back into the air. What we need ideally is a vacuum bag with a pore size of 0.5 micron or smaller. Larger particles cannot get through the little holes in the bag. They will be trapped. Some solutions:

HEPA Vacs have a HEPA filter on the exhaust so that no particles larger than 0.3 micron can escape back into the air. They are more expensive than ordinary vacuums because they need a larger, more powerful motor to suck sufficient air through those tiny holes. If you buy one, make sure the filter is a ***true*** HEPA and not a HEPA-like, a HEPA-style, or a 95 percent HEPA filter. A true HEPA filter will specify that it "removes 99.97 percent of all particles that are 0.3 micron or larger." If those words aren't used, then it may or may not be a true HEPA. You may not be trapping those tiny allergens inside the bag, despite the premium cost of the equipment.

Water-capture vacuums can be quite effective because water has no pore size. If properly designed, they theoretically can remove all particles of any size. They still should have a filter on the exhaust because not everything is water soluble. Drywall dust, pepper, plaster, wood ash from fire places, and talcum powder, for example, usually will blow right through the water.

Special allergy bags and micro-filtration bags now are available but are not much help unless the pore size is 1.0 micron or less. Standard replacement bags have an average pore size of 8-10 microns. The special allergy, or micro-filtration, bags have an average pore size of 5 microns. That means almost everything less than 5.0 microns will *not* be trapped inside the bag. Particles will blow through the bag and back into the air. These bags are an improvement, but usually not enough. One exception is the 1.0-micron-pore size bags made from DuPont Hysurf® material. These have proven extremely effective, even for people with extreme susceptibilities.

Be wary when reading labels and specifications for vacuum cleaners and vacuum bags. The wording can be very misleading. For example, it is not uncommon to see the following statements for vacuum bags:

- Removes over 99 percent of common allergens.
- Removes particles as small as 0.01 micron.

I can't tell you how many people have called to say they just found a great vacuum bag that is better than a HEPA vac for less than $2.00 a bag. When I ask why they think that is true, they respond, "It says right on the label that it removes over 99 percent of 0.01-micron particles. HEPA only goes down to 0.3 micron. This is 30 times better!"

They are quite disappointed when they realize that what they see and what they get are too very different things. First, the label does not claim to remove 99 percent of 0.01-micron particles. It does not make one statement making both claims. It makes two different statements, with two separate claims:

1. The first claim is that it removes 99 percent of common allergens. That statement will be true if allergen removal is measured by ***weight***. Obviously, the large visible particles from 10 to 100 microns will weigh much more than the pesky irritants below 10 microns. A 10-micron bag can remove several pounds of large, heavy allergens and still not capture many of the tiny ones below 10 microns.

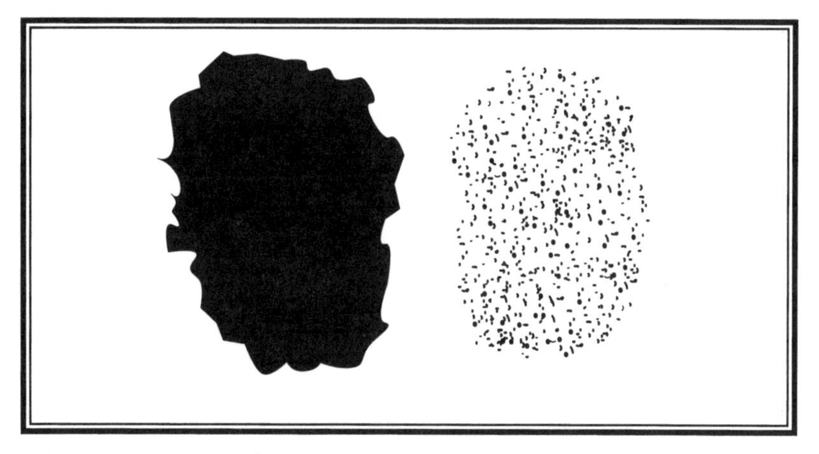

A large 100-micron-diameter particle could easily weigh 1,0000 times more than a whole cluster of particles that are only 1 micron in diameter. When evaluating the particle removal effectiveness of vacuum bags and air purifiers, it is critical to know the difference between ratings based on weight and ratings based on size.

2. The second claim is that it removes particles as small as 0.01 micron. That statement is technically true if only ***one*** particle of that size is removed. Even screen wire will catch an occasional piece of microscopic dust.

The claims *appear* to be great, based on the information they present. However, what they leave out makes them misleading.

The only reliable way to know what you are getting with vacuum cleaner bags is to know its pore size. That is the *only* description you can trust. Buy a HEPA-vac if you can afford it. They have a pore size of 0.3 microns and cost $800 and up.

Some lower cost vacuums that have HEPA filters sell for about $200. They meet the technical definition of a vacuum with a HEPA filter. But don't expect them to clean the carpet and filter the exhaust as well as the higher cost ones. They are sort of like riding a bicycle instead of driving a sports car. Yet they meet the needs of the majority of all but those with the higher PIRs.

Others are labeled as HEPA-type or HEPA-like. If they don't have a HEPA filter that has specifications that include the words "99.97% removal of particles 0.3 microns and larger." That means their HEPA-like isn't a HEPA filter. They usually have a bag and filter system no more effective than most other vacuums.

An elegantly simple solution to all this is to buy a central vacuuming system. There are no problems with pore size and retention of differing size particles as long as the vacuum exhaust is outside habitable areas. The garage usually works fine, but I have seen them installed in the laundry room, right next to a home-office space, and even in a closet in the master bedroom!

Our goal with using vacuum cleaners is not to just "vacuum the rug or upholstery." The goal is to effectively remove the source of an undesirable source of exposure, without merely changing its location or replacing it with another reactive agent.

AIR PURIFICATION DEVICES

Although room purifiers and furnace filters *are* designed to remove subvisible particles, their special twist is the variety of industry standards that exist for the testing of air filtration devices. There are six of them. Plus a consumer standard whose intent is to measure practical effectiveness, despite all the technical jargon and posturing.

Manufacturers of filter media want test results that give them an advantage over their competition. That's understandable. And their claim that different techniques of removing pollutants require different testing methods is also valid. But the end result to the user is even more confusing than that of vacuum cleaner bags.

Some testing standards measure particle removal by weight, others by particle count. Some measure particle count, but only of particles larger than 10 microns. Others use a mix of particles from 0.5 micron to 100 microns. Some measure the effectiveness only after the filter is dirty, when a lot of the pores already are clogged with dirt. Electronic filters are tested only when the dirt is first introduced to the filter because the effectiveness deteriorates with use.

Another complicating factor is the way some filter media manufacturers increase the effectiveness without reducing pore size. They add an oily substance or something sticky to the media so that some particles smaller than the pore size will stick to the filter media and give a better test rating. However, chemically susceptible people often react to those substances.

Self-charging electrostatic filters are a variation of this technique. But, instead of a sticky chemical, electrostatic charges increase the effectiveness of their filtration. However, humid climates tend to drastically reduce the amount of static charge, thereby reducing effectiveness. A commonly occurring oily film from the air also reduces the static charge and must be removed periodically with detergents. Even so, they are a substantial improvement over standard fiberglass matt filters — which , by the way, are intended more to protect the equipment from large buildups of dust and debris rather then to meets the needs of the human respiratory system.

Cost of electrostatic filters can range from $40 to $200, depending on size and rating. Typically, the $40 to $60 filters have an Efficiency Rating (one of the six standards for rating filters, not to be confused with the rating method for HEPA, electronic or other types of filters) of about 50-60 percent. The higher cost filters, those typically above $100, should have an Efficiency Rating of over 90 percent.

One of the better overall filters on the market is a disposable filter that also functions, at least partially, on electrostatic charges. It is the one with white pleated material whose description includes the word "filtret" or "filtrete." (Filtrete® is made by 3M Company.) They cost anywhere from about $14 to $18 and must be replaced more frequently than standard type filters to avoid unacceptable reduction of airflow due to clogging of the much smaller pores. They are more effective than many of the electrostatic filters but the long-term cost of the 3 to 8 changes per year can easily exceed the $100 to $200 initial cost of the upper-end electrostatic filters.

Electronic filters are excellent. They remove particles by passing the air through a high voltage grid. It literally "dust plates" the filter grid like you would "chrome-plate" an (old style!) car bumper. Although it has no pore size, it is extremely effective at removing the smallest of particles, even those below the 0.3 microns of the HEPA filter.

However, you must be aware of its limitations. Unlike filters that strain the particles out of the air, which actually become more effective as they get dirtier because the pores become clogged, electronic filters become *less* effective as they get dirtier. The build-up of dust, dirt and oil-film tends to reduce the electrical charge on the plates of the grid, gradually reducing its ability to attract and hold additional particles. The dirtier it gets, the more it lets through. Public utility companies typically suggest that the collection grids be cleaned at least once a month. People with a high PIR and an awareness of subtle changes in exposure, suggest cleaning every 3-4 weeks.

The "snapping" sound of electronic filters is okay only when it is first turned on to let you know it is working. If the "snapping" continues the filter is either malfunctioning or debris is shorting the grid. The "zap and the snap" does not mean that the filter is cleaning itself. On the contrary, each time the grid shorts as the electrical arc "snaps," the high voltage field temporarily collapses, allowing everything to pass through.

Electronic filters can generate slight levels of ozone. This usually isn't a problem unless you are highly susceptible to ozone. A few of my clients cannot tolerate electronic filters for this reason. In fact, they also have trouble with the ozone generated by the slight electrical arcing inside the motors of furnaces and refrigerators.

Ideally, evaluating filtration media according to pore size is the best. The HEPA filter is the only one that uses that rating technique. Its rating is

99.97 percent of particles 0.3 micron and larger. However, the HEPA is not always feasible. The air resistance is so high due to the extremely small pore size that it requires a large motor. This leads to more size, cost and noise.

Finally, HEPA filters cannot be used as a furnace filter unless either the furnace blower is specifically designed for it or the HEPA filter has its own blower. There are a couple of sources for these, with costs starting at about $900, depending on size, plus installation.

HEPA furnace filter like this one requires its own blower to overcome the large air resistance. The whole unit pulls air from the return and through the filters. It then returns the filtered air to the furnace just before the regular blower.

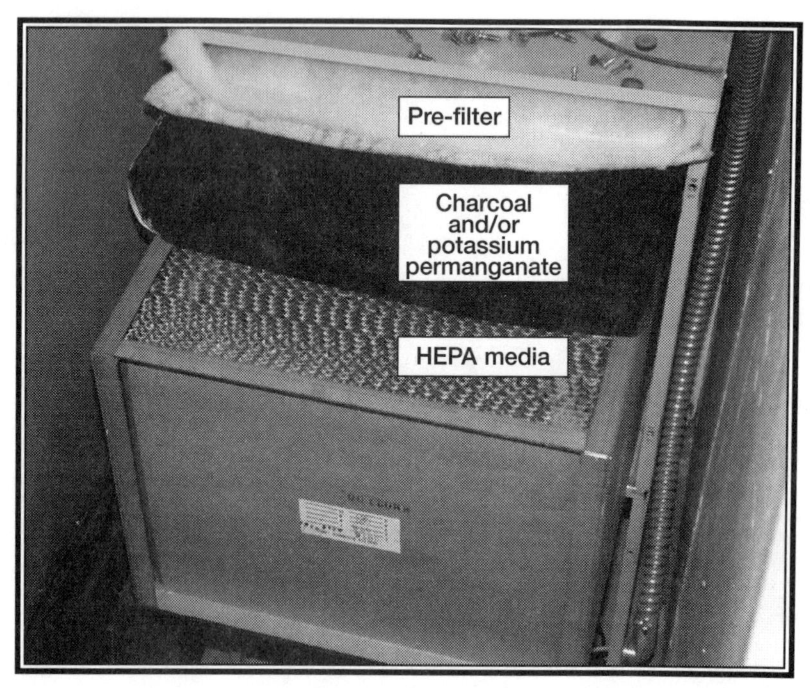

The most effective and cost effeciant HEPA furnace filters have three stages. The pre-filter removes the larger particles so they don't clog the expensive HEPA media. The charcoal and/or potassium permanganate removes odors and chemicals. Finally, the actual HEPA media, usually measuring at least 12"x12"x12", removes the subvisible particles as small as 0.3 microns.

Although a HEPA filter is one of the best types and has reliable specifications, other types can be good enough for most people — especially those with low PIRs. Try them and see what benefit you derive and make sure you donn't react to the purifying device itself. Also, buy from a company that allows a money-back return policy so you don't end up with a house full of useless gadgets and equipment.

Removal of chemicals, odors and fumes by air purification devices present an entirely different problem. Chemicals are not removed by the same methods as particles. They pass right through most filters like dust through screen wire, or like allergens through standard vacuum cleaner bags.

Chemical removal requires a filtration media that the molecules will either adhere to or chemically react with. Materials such as activated carbon, potassium permanganate and zeolite are the most common. Several key points are:

The most effective chemical removal material is determined by the chemicals you want to remove. Each of the three materials has properties that end themselves more to one type than another. Charcoal is particularly adept at removing petroleum-based products such as auto emissions. However, it is not as effective with the one chemical that affects most sensitive people: formaldehyde. Potassium permanganate does a tremendously better job with that. Finally, zeolite has a much higher affinity for ammonia-based substances, such as animal urine. Most manufacturers of high-quality purification devices offer a mixture of two, or even all three, media, in one package.

The more media, the longer it will last. Charcoal pads are quite inexpensive, compared to granules of charcoal. But they don't last as long as a tray full of granulated charcoal and must be replaced more frequently. A pad containing a few ounces of powder may have to be replaced once a week while a tray containing five pounds of granules may last a year. The pads, although costing less at first, usually cost more in the long run.

Educate yourself on the proper use, maintenance and replacement of the chemical media types. All three require different actions. When the charcoal is full of chemicals, not only will new chemicals pass through, but the ones already trapped may be released back into the room. The charcoal must be replaced. Potassium permanganate, on the other hand, chemically reacts with what it comes into contact with. It oxidizes it. When the pellets change color they must be replaced. However, sunlight or heat can "recharge" zeolite, which releases the ammonia products that have been adsorbed. Non-ammonia substances, such as fragrances, usually will not release. In that case the zeolite must be replaced.

An entirely different way of removing particles, and occasionally chemicals, is with *ionizers*. Negative ion generators produce large amounts of negatively charged ions that attach to particles. The newly charged particles are then attracted to surfaces that normally have a slight positive charge, such as floors, ceilings and walls. If the negative ion field is strong enough, the ionizer can be quite effective at removing particles. In fact, ionizers typically are most effective with the smaller particles, those below the 0.3 micron limit of HEPA filters and other mechanical purification devices.

Also, they are not limited by the "gate" effect of purifiers that have blowers. The blower types remove only those pollutants present in the air that actually passes through the purifier. It has little, if any, effect on the far corners of a room and will not remove anything from the walls, ceilings, upholstery and other surfaces. In fact, if there is a lot of "dust" in the house, the blowers often can increase the complaints for awhile by creating a different air movement pattern, which stirs up the dust and redistributes elsewhere. Ionizers, on the other hand, neither create nor disturb air patterns.

Ionizers occasionally can reduce chemical sources, especially those that don't respond to normal methods. Chemical molecules are thousands of times smaller than even a 0.01-micron particle. These molecules often adhere to the particles, much like they do to charcoal filtration media. Under those conditions, airing out the house has little effect because the chemical location and movement is now governed by the characteristics of the particle it is "hitchhiking" on. As the ionizer moves the particles out of the air and onto surfaces, the chemicals go with it. This explanation also helps to explain why HEPA filtration sometimes appears to help reduce chemical sources.

I had to make three trips to various stores and place two catalog orders before I could find something that would clean the air sufficiently for me and that I wouldn't react to.

I wanted a HEPA filter but most of the ones I could find used lots of plastic. I reacted to all of them. Even one that was mostly metal had an unacceptable odor coming from the motor.

Most "HEPA-like" filters just were inadequate. But there was one from a mail-order supply house that worked just fine — but only in the bedroom.

What's curious is that the same filter wouldn't work in my sewing room. Only a negative ionizer worked there.

Why the difference? I don't care. I solved my problem and that's all that matters.

All filtration techniques have their strengths and their weaknesses. Some will work great in your application and some won't. If you don't know how to determine which is best, just pick one and try it. If it works, terrific. If it doesn't, take it back and try something else. Just make sure you buy your product from a source that will let you return it, satisfaction guaranteed, no questions asked. That's how you protect yourself during the frustration of the trial-and-error process.

Again, the goal is not to "use an air purifier." The goal is to remove whatever is in the air that is the source of an undesirable exposure, without replacing it with another problem agent.

CLEANING PRODUCTS

By their very nature, cleaning products are usually chemicals. It is critical that you understand your PIR for chemicals in general, and even for specific chemical families. For example, if ammonia gives you headaches, but you have no problem with a petroleum distillate-based solvent, then use the solvent and avoid the ammonia. If they all cause problems, then you have to be very vigilant. The chemical families that commonly cause concern are:

- **Ammonia**
- **Petroleum distillates**
- **Chlorinated compounds**
- **Alcohol and alcohol-based compounds**

So what's left? Numerous cleaning products are now available that avoid these ingredients. Look for the ones that are vegetable-based. If fragrances cause problems, find ones that either have a fragrance you are okay with or that have none at all. You often can make your own from basic materials such as vinegar, borax and baking soda. (See Appendix B)

But what if you can't select the product for your own use? How do you know what is in the cleaning agent the carpet cleaner or the weekly cleaning service uses?

Start with the label. Read it and reject ANY product that you are not sure of. If the label looks all right, then request the Material Safety Data Sheet (MSDS). The MSDS is available for most cleaning products. If the government requires that the products' manufacturer make it available to their workers, they must also make it available to anyone who requests it. Obtaining one usually is quite simple. If the local store or distributor doesn't have it, you can call the manufacturer yourself. They usually can fax it to you in less than an hour.

Again, the goal is not to "clean something." The goal is to remove whatever is in the house that is the source of an undesirable exposure, without replacing it with another problem agent.

DISINFECTANTS AND PESTICIDES

Disinfectants and pesticides are designed to kill living things such as bacteria, mold, insects and weeds. And that makes them more of a concern than cleaning products because many of these are suspected of being neurotoxic to people. Others haven't even been tested so no one knows if they are safe.

Regulatory agencies assume that a deadly dose for tiny life forms, such as bugs, weeds and germs, is not large enough to kill people. The principle of killing by chemicals is that the size of a lethal dose is dependent upon the body weight of the target. Because bugs and weeds are much smaller than people, a minimum lethal dose for them is not nearly enough to kill us. And that is true.

But what is the effect on humans of a continuing sequence of small doses? Some substances accumulate in body fat, meaning that we may eventually be exposed from inside our own bodies. Noted environmentalist Rachel Carson first informed us of this in wildlife. As you move up the food chain, the higher levels of life tended to have higher quantities of pesticides stored in their body fat. This also applies to humans.

There also are questions about the mixing of chemicals. What happens when several disinfectants and pesticides are mixed together? One common industry technique is to apply a mixture of fertilizer, insecticide and herbicide. That single application of all three chemicals makes good economic sense. But what are the effects of that *mixture* on people? No one has tested the mixture. What new chemical has been created? What are the short-term and long-term effects? And what effect, if any, do they have on our bodies?

Present research is inadequate to answer any of these questions. The vast majority of EPA registered pesticides have not been tested specifically for neurotoxicity or cancer in humans. Other non-lethal effects, which are the main concern of this book, typically are ignored as anomalies.

It is not my intention to recommend that all pesticides and disinfectants should be banned. Nor is it my intention to claim that their potential for danger to humans outweighs the obvious benefit of their proper use. My concern is that the current debate about chemical effects focuses on a statistical public at the expense of *specific* individuals.

Remember, these products are governed by laws that are based on testing for *Public* Safety, which are based on statistically calculated toxicity curves. I know of no studies or laws that even consider the other two types of reactivity curves, atopy and sensitivity. Furthermore, none say anything at all about any specific individual. None of the standards, cautions, approvals, or claims say anything about ***you***.

Inform yourself by reading labels, the MSDS, and any other sources of information you can find. Make your own choices and be responsible for yourself by relying on your own expertise rather than on the authority of others.

Theoretically, we should have no problem with being responsible for our own safety. However, we are prevented from doing so because current laws and regulations don't compel businesses to provide us with accurate and complete information. They are required only to follow specified instructions to meet the minimum legal requirements of labeling laws. Therefore, you cannot rely on the manufacturers, contractors or applicators of disinfectants and pesticides to make accurate and reliable decisions concerning *your* safety. They don't know. They can't know. But *you* still have to choose, whether the information is complete and accurate or not.

The goal is not just to "kill the bugs." The goal is to kill whatever living organisms are the source of an undesirable exposure, without replacing them with another problem agent.

DUCT-CLEANING SERVICES

You can't do this job yourself. Sticking the nozzle of your vacuum cleaner down the vent as far as you can reach is inadequate because the dirt of most concern is deep inside the trunk lines. I've seen material two inches thick on the bottom surface of the ducts. The "dirt" is great food for mold, bacteria and other living creatures. Even the thin film of dust on the top and side walls is important because it often contains mold and other microorganisms. In fact, the whole duct system is often an ideal habitat for a variety of problems. It's dark, dirty and often damp due to either natural humidity or central humidifiers. What is visible is often just the tip of the iceberg.

You must have a professional service company clean the forced-air ducts. Only they have the appropriate equipment and knowledge of how to use it to actually remove foreign materials from the system.

When you select such a company, consider the following requirements:

Sufficient airflow is necessary to move the debris from the ducts into the collection source. Large truck units with 8-inch or larger hoses usually meet this requirement. However, some small "shop vac" devices also can be sufficient if used properly; the end of the vacuum hose must be moved through all the ducting. That is the only way to achieve adequate airflow to remove the debris.

A large vaccum hose is required to create sufficient air flow thru the entire system. Hoses smaller than 8 to 10 inches cannot do that. They must be "snaked" thru the ducts, end to end.

Airflow is not enough. Devices that physically contact the duct surfaces, such as brushes and rotary augers, are necessary to physically loosen and remove the debris and the thin film of "dust" from the sides, bottom, and top surfaces of the ducts. If airflow were all that was necessary to clean, we would never have to wash our car, for example. We'd just drive it 100 miles per hour down the freeway.

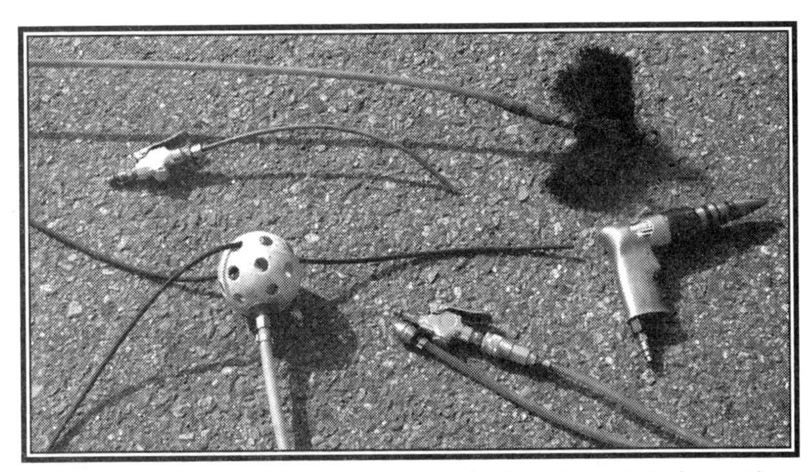

A variety of tools are necessary to properly clean forced air ducts. These include high pressure air nozzles, "skipper" balls, manual brushes and a variety of rotary brushes.

The vacuum source should be outside the house. That is the only way of preventing the contents of the ducts from being redistributed throughout the inside of your house, just like typical vacuum cleaners do with carpet dust. Some service companies claim that the HEPA filters on the equipment exhaust will block *all* dirt. The truth is that a HEPA, or any other similar type of filter, will block only those particles larger than the pore size of the filter media. Is the HEPA the best filter? Probably. Will it remove everything? Absolutely not. There are many particles smaller than the 0.3-micron pore size of a HEPA filter. So be safe, especially if your PIR is above 3. Insist that the vacuum source be outdoors. If the service refuses to accommodate your needs, refuse to do business with them.

At the other end of the particle size spectrum, don't be overly impressed by companies that emphasize the presence of lumber, toys, bricks, sheet rock, nails and other large objects that they find in the ducts. These objects aren't of major concern because they are too large to inhale. Rather, find a company that emphasizes the removal of the tiny, subvisible particles typically found in the thin film of dust on the duct walls. Those are the particles that you are most likely to breathe.

Companies also love to make claims about the importance of removing dust mites from the ducting. Ignore the scare tactics and the gruesome photos of the mites. Dust mites, although a serious allergen and respiratory irritant, are rarely a primary problem in forced-air systems. Even if the mites are alive and thriving, their feces, which is the problem, are typically too heavy to stay airborne for any substantial length of time, dropping to the floor within a foot or so of the heat vent. When you should be concerned, though, is if any supply vents are located near the top of the wall or in the ceiling — especially if that is directly above where you sit or sleep.

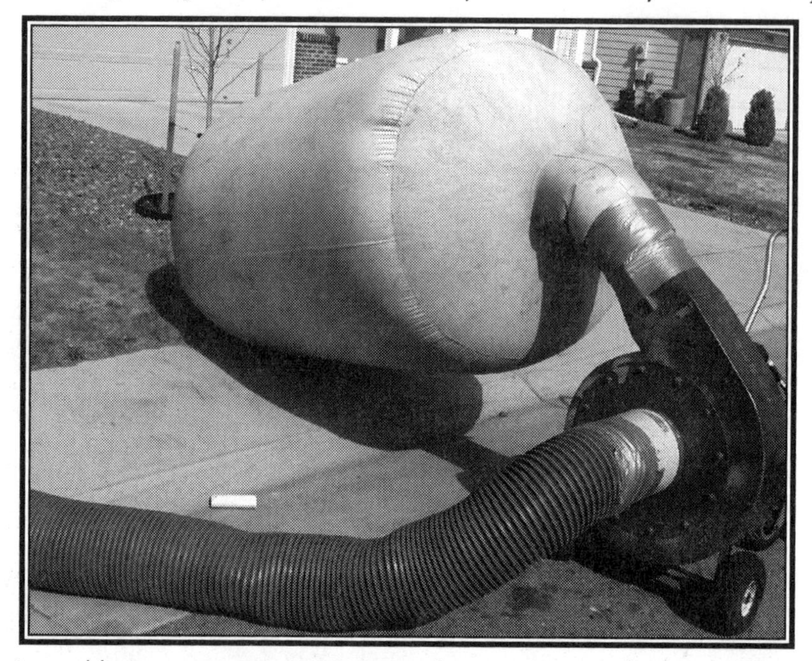

Large blowers are necessary to move large amounts of air. Even larger ones are inside truck mounted units.

Pet dander, especially cat dander, is a major concern. Because it can remain airborne for prolonged periods of times, it travels wherever the air goes, which is everywhere. If pet dander is not removed from the ducts, it will continue to blow throughout the house, continuing your exposure to it. Choose a company that understands the characteristics of pet dander.

Mold is of special concern, because that source is often replaced with a chemical source. Most cleaning companies insist on using a disinfectant to ensure that all the mold and other microorganisms have been killed. That is important. But be careful what disinfectant you allow them to use. Read the label and MSDS. Test it yourself. You decide. Do not rely on anyone else to make this decision. If they are reluctant to meet your special needs, find another company.

Another important criterion for choosing a duct-cleaning service is their attitude about the use of sealants inside the ducting. Sealants often are recommended to ensure that whatever dirt is missed cannot reenter the airflow. The sealant will "glue," or encapsulate, the dirt to the duct walls. Again, treat it as you would any other chemical source, with the added proviso that once applied, it cannot be removed. Determine your PIR and then read both the label and the MSDS. Beware of the fragrance! Then make your decision. However, if a cleaning company does a proper job of cleaning the ducts, why would any debris be left to be sealed in?

The furnace blower cannot be thoroughly cleaned unless it is first disassembled and removed. The blower is critical because all the air must go through the blower. If the blades are dirty then the air, even if it has just been filtered, will be contaminated. If you are chemically sensitive, beware of solvents and lubricants that may be used.

Again, the goal is not to "clean the ducts." The goal is to remove whatever is in the ducts that is the source of an undesirable exposure, without redistributing it or replacing it with another problem agent.

CARPET CLEANING CHEMICALS AND SERVICES

Carpet cleaning is a job you *can* do yourself. Equipment usually is available for rental or purchase that will, if used properly, adequately clean your carpets and upholstery.

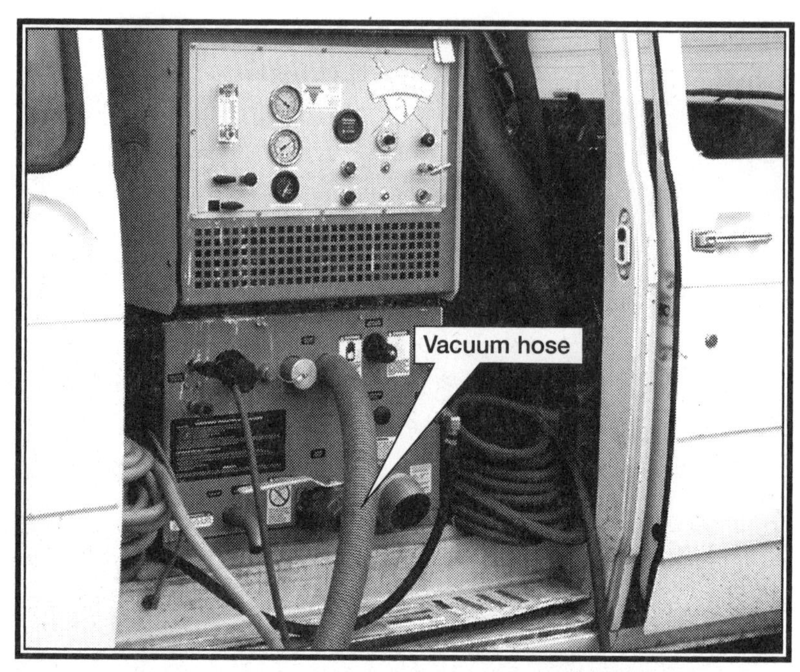

Truck mounted carpet cleaning equipment provides the hottest water and the most air flow for removal of the water and the cleaning solution. Notice the size of the hose. It's diameter is fine for carpet cleaning but if it is also used for duct cleaning, special techniques and additional tools are necessary for an adequate job.

Whether you do the job yourself or have a professional company do it for you, the type of equipment is important. I recommend only hot water extraction. It is the only method I know of that applies and then removes the cleaning solution in large quantities. If the cleaning solution isn't extracted from the carpet, then the dirt will not be removed either.

Some techniques clean the surface fibers very well but do not remove the dirt. The dirt is pushed deeper into the carpet. Companies that use those techniques claim that their methods release the dirt from the fibers, making them available for removal by routine vacuuming. If the fallacy of that isn't readily apparent, go back and read the section of this chapter on vacuum cleaners.

The cleaning solution itself and the fragrance it typically contains often cause more problems than the dirty carpet. Read the label and the MSDS. Test it. Make your own choice rather than rely on the cleaning people to tell you it is safe. And beware of the residue it is possible to leave behind in the fabric!

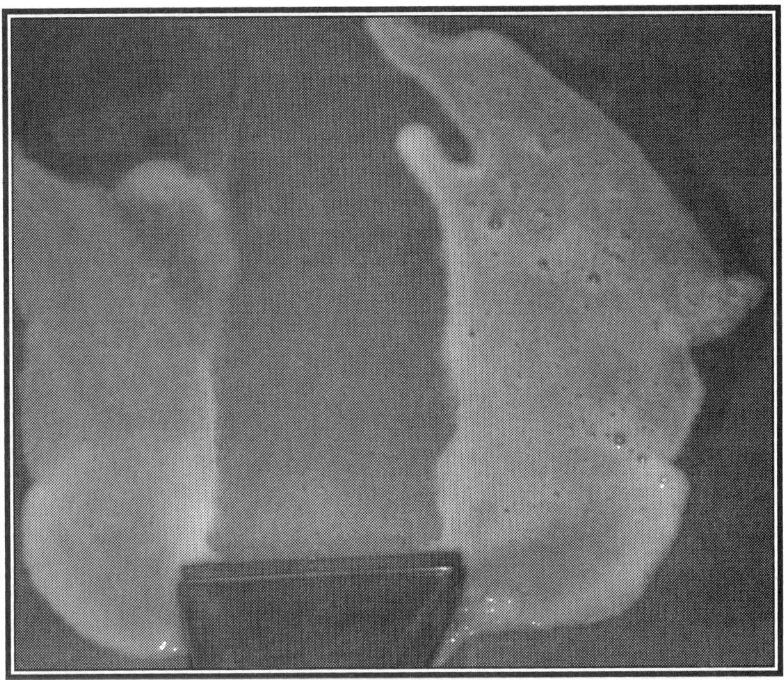

A serious problem can occur with improper use and inadequate removal of the cleaning agent. This photo shows the suds after a professional company had just cleaned this couch. If the cleaning agent also has a fragrance or other odor, the couch now becomes a source of chemical exposure.

If you are renting the equipment, you may need to rinse the previous cleaning agent out of the tank and hoses. That may take 5-6 tanks of water, depending on what was previously used in it and what your PIR rating is for chemicals.

I usually do not recommend any carpet-cleaning process that uses dry-cleaning chemicals. They are not designed to remove the cleaning solution in sufficient quantities to also remove the dirt. They also typically replace a dirt problem with a chemical one, especially for those with a high PIR for chemicals.

I also don't recommend stain-resistant chemicals for either carpets or upholstery. Such chemicals are common triggers for some of those with a high PIR for chemicals. Chemicals applied after the carpet is manufactured usually don't work very well or last very long. They need to be added at the factory so the fibers can be sufficiently saturated and coated. Also, the benefit is somewhat illusory. They don't prevent the presence of dirt. They merely make the fibers slick enough so the dirt falls deeper into the carpet. The carpet looks better but the dirt is still there — you just can't see it.

Again, the goal is not to "clean the carpet." The goal is to remove whatever is in the carpet that is the source of an undesirable exposure, without redistributing it or replacing it with another problem agent.

TEST THE PRODUCTS

Before you or your service provider is allowed to use any product, test it for your individual safety. Begin with the label and then the MSDS. If those seem okay, then personally test the product. Smell it and touch it. If you react to a brief exposure then how can you ever tolerate a house full of it? If it does pass this initial exposure test, then take a small sample and set it near you for a longer exposure. If you can tolerate it for several minutes, then you might consider placing it by your bed and increase your exposure to all night. Again, if you react within any of these relatively short time-frames, how are you going to tolerate living with it? If it does pass these tests, then you know that the risk of reacting to it has been reduced. Remember to factor in your PIR. The higher the rating the more careful you must be on exposing yourself, even on a trial basis, and the greater the risk that you will react later to the larger exposures.

TEST THE RESULTS

The ultimate test for the success or failure of the cleanup is the change in your symptoms. Notice which symptoms are better and which are worse. As you gain more experience with which sources are correlated with which symptoms, you will increase your ability to evaluate the effectiveness of products and services. Use companies that are cooperative with your needs. Try to buy from companies that offer a guarantee of satisfaction so you can return anything that doesn't help *you*. That allows you to experiment without being stuck with lots of unusable equipment.

ALLERGEN TESTING

A "settled dust" method of testing is designed to detect allergens from cockroach, dust mite, cat dander, dog dander, mouse urine and mold. Sample collection is simple and can be done by yourself using your own vacuum cleaner.

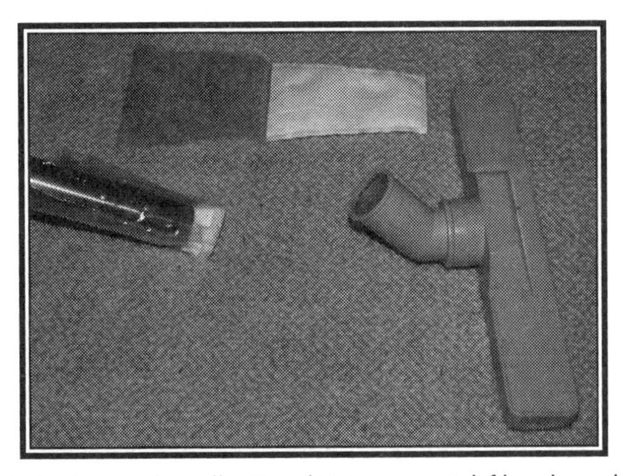

The settled-dust sample collection kit is a special filter bag designed to be used with an ordinarry vacuum cleaner.

The sample collection kits and subsequent analysis services are available only by doctor prescription from The Johns Hopkins University, Reference Laboratory for Dermatology, Allergy & Clinical Immunology (DACI). Ask your doctor to obtain the necessary information about price, sample collection instructions, and medical results evaluation by writing to P.O. Box 9975, Baltimore, MD 21224. Or call **1 (800) 344-3224**.

A commercial lab providing the same analysis technique, available to either your doctor or an IAQ contractor, consultant or service provider, is available from Indoor Biotechnologies Inc., 1216 Harris Street, Charlottesville, VA. Or call **1 (804) 984-2304.**

This testing protocol can also be used to help you evaluate the success of your source removal attempts. Collect a sample prior to any work to establish the level of exposure. Then collect another sample after the work to determine the change. Even if the final result isn't zero, which is often typical, you may be able to determine the level of exposure that triggers your reactions. If your symptoms for various sources are unique to that allergen, you can eventually determine your exposure based on your symptom.

This procedure is the only objective measure for allergen removal that I am aware of. Unfortunately, it does not test for everything. For example, it won't detect chemicals, non-fungal microorganisms, and nonallergenic particles. Therefore, if you have symptoms from problem agents that this test doesn't detect, you will still have symptoms despite negative test results.

Chapter Eleven

Evaluating a House

Evaluating a House

The "common sense" starting point for evaluating a house for indoor irritants, allergens and asthma triggers is to simply buy an air purifier. But if that doesn't work, a more effective approach is to locate sources and then remove them. That works in the vast majority of situations. But if it, too, fails to successfully stop your complaints, you may need to change your starting point by making the following adjustments to your approach and to your assumptions:

- Assume the house *is* the cause.
- Change the interactions of sources with each other and with the occupants.
- Consider what you *haven't* done.
- Examine the timing of the complaint relative to the exposure.

ASSUME THE HOUSE IS THE CAUSE

Typically, we start with the assumption that nothing is wrong unless proven otherwise. We assume that the house is innocent until proven guilty. That approach is fine for a mature science where there are laboratory tests and the causes, effects, and relationships are well known. But that approach is exactly what does *not* detect the new types of undifferentiated problems you are most likely struggling with.

Rather, start by assuming that there ***is*** a problem in the house. Find as many potential problems as you can, whether its only one or one hundred. Then, rank them as to which are most likely and which are least likely to cause your complaints. Identify, if possible, specific symptoms with types of sources and possible locations. This is important, because as you make changes in the house, you will monitor and experience differences in your symptoms as these changes are accomplished.

My house can't be the problem. I had the furnace checked and the inspector said it's normal to have a slight natural gas smell in the furnace room. He said that the ducts were dirty but that is typical also. He then pointed out that the combustion make-up air was coming from the adjacent crawl space. Although that's not to code anymore, it was okay when the house was built.

Besides, we got rid of the cat last week and bought an air purifier that's as good as a HEPA but costs a lot less money. We also installed several of those plug-in deodorizers so we don't smell the musty basement anymore.

My wife and I are still just as sick as before, but everyone we asked said the house isn't the problem.

CHANGE THE INTERACTIONS OF SOURCES AND OCCUPANTS

Not all sources of exposure are specific locations of particles, chemicals or living organisms. Sometimes the sum total of "whatever" is there is most important. Because a habitat is the interactions of all components, the mere presence of mold *or* chemicals may not be sufficient to cause your discomfort. But the combination of the mold with the chemicals may be. The problem could intensify in the winter when the windows are closed, allowing all sources to accumulate. Add to this the forced-air heating system, which may increase the circulation and distribution of the sources.

Other examples include a damp crawl space, evaporative cooler, a very old house and insulation inside forced-air ducts. Insulation is a common reservoir and amplifier of living organisms, which are then distributed throughout the house by the forced-air system. Other sources are the insulation itself, the glue that holds the insulation together, and another glue that sticks it to the duct wall. Furthermore, the insulation tends to collect and hold dust, creating an even more favorable environment for mold. All that, combined with the mold source in the damp crawl space and the evaporative cooler, plus the microbial activity in the deterioration of an old house, and you have the potential for very complex combinations of sources.

To further complicate matters, if your main symptoms are associated with chemical exposure, you will tend to ignore the mold and the conditions that promote mold growth. However, mold generates chemicals — VOCs. Large quantities of mold actually may be the *primary* source of chemical exposure. The best response to your complaints may be to reduce the mold levels, thereby reducing the chemical exposures. This requires several actions: removal of the mold sources, such as the insulation inside the ducts; isolation of the mold sources in the crawl space and evaporative cooler; venting the crawl space so the mold won't continue to thrive; closing all wall and floor penetrations to isolate the mold inside the old walls and floors; and, finally, a one-time ozonation treatment to kill the persistent airborne mold.

WHAT HAVEN'T YOU DONE?

This is fairly simple to understand but difficult to execute. The simple part is that whatever you *have* done hasn't solved the problem. So what have you missed? Figuring that out can be like trying to think of something you've never heard of before. It's like trying to read somebody's mind.

To begin this process, look for *anything*, not a specific object. If you only look for a needle in a haystack, for example, you will miss thousands of other objects that may be important. Instead, look for anything that you can find in the haystack, including the hay itself. Now you will notice thousands of objects, any one of which may be important.

If you decide ahead of time what the problem is and what it isn't, you may miss the real one(s).

Here's an example: Your new cleaning lady, against your instructions, used ammonia on your kitchen floor tiles. You immediately have her mop again with your "safe" detergent, but an offensive odor of ammonia still lingers. So you have her mop the floor again — but with the same results.

The assumption is that she hasn't done a proper job of cleaning. So you have her mop the floor again. And again. And again. And still the odor lingers. You are focusing too intently on what you think is the problem. Instead, look for a different reason for why the ammonia odor is not being removed.

The solution? The floor is old and has cracks between the tiles. The dirt in the cracks has absorbed some of the ammonia. Remove the dirt from between the tiles and the ammonia goes with it.

If you find yourself doing the same task over and over, stop and consider what you are *not* seeing about the situation. You are missing something.

The usual reason we look for the wrong needle in the wrong haystack is that we make certain assumptions. If those assumptions are true, you will be successful at stopping your exposures. However, if you are not successful, those assumptions most likely are wrong.

Here are some examples of false assumptions that interfere with success:

- If something is "normal," it can't be a problem. Deadly radiation at ground zero of a nuclear bomb blast is "normal," but you wouldn't want to be there during or after the blast. It is normal for a crawl space to be wet and smelly, but you wouldn't want to live there. And it is equally "normal" for crawl space odor and mold to travel on air currents to *everywhere* in the house.
- Just because it's natural doesn't mean it's "safe." Arsenic is natural.
- Just because it's plastic doesn't mean it's "bad." A lot of plastics are sources of chemical exposures, but some are less so than stainless steel.
- Just because a chemical is inert does not mean it is safe. Asbestos is chemically inert, but it may still be a cancer-causing substance in certain situations.
- Just because something is "odorless" doesn't mean it has no fragrance. There are "neutral" fragrances and there are "neutralizing" fragrances.
- Just because it is widely available at the local mall doesn't mean it is safe *for you.*
- Just because the federal government says it is safe doesn't mean *you* won't have an adverse reaction.
- Just because the person installing the carpet pad or running the duct-cleaning or carpet-cleaning equipment says his techniques are safe doesn't mean they are safe *for you*. And don't rely on the assurances of their managers, either.
- A common statement by people in any industry is that their products must be safe because they have been working personally with them daily for years and have never been sick from them. That is probably true. However, if they did not have an unusually high resistance to what they work with, they wouldn't have stayed in the business. They would have felt ill or been otherwise unhappy — whether they associated their evaluation with environmental exposures or not. But that doesn't mean the product is safe for them — or *for you.*

- Removing the cat or dog will stop the reactions to its dander. Not so. The *source* of the dander has been removed, but the *accumulated* dander is still there.

- Covering the dirt in a damp crawl space with plastic is assumed to be the best way to stop mold. Actually, it will decrease the amount of moisture in the air of the crawl space, and for several years it actually may help to reduce or prevent mold exposure. But this only slows down an inevitable process. The moisture in the dirt is held in by the plastic, allowing it to gradually accumulate; or even become muddy as in this photo. And now the conditions under the plastic may be ideal for enhancing the growth of mold and other microorganisms.

Unless all the seams and the perimeter of the plastic are sealed, creating a complete vapor barrier as is done for radon mitigation, the increased mold population eventually will escape into the air. The problem is now more serious than before. Venting the moisture rapidly out of the bare dirt is a more effective remedy. However, not having a crawl space to begin with is ideal.

TIMING

The concept of timing is not a new starting point. Rather, it is the act of combining the starting points you have already established for your susceptibilities with those for your exposures.

It pinpoints the time connection between an event that creates or amplifies a source and the onset of your symptoms.

For example, there is almost always a connection between when you first got sick and when you moved into your present house, or when the basement flooded, or when you started letting the dog into the house — and allowing it to sleep on the bed.

Solving our house problems took quite awhile. We've lived here for over 20 years and never noticed any change until we built an addition three years ago. He got sick and just never seemed to recover. I thought it was either old age or he had become allergic to me!

But then we noticed that when we went on vacation we both felt better. Then when we came back home and first walked into the house, it smelled. I can't really say what it smelled like but it smelled. By the next morning we didn't notice it anymore.

Last fall we replaced all the carpet in the basement and the smell went away — and so did most of his symptoms.

But then I had problems. I couldn't stand the odor of the carpet. Actually it was the pad that stank, like car tires. We had it replaced and now I'm better.

We got to talking the other night and realized that the basement carpet actually started smelling after it got wet when the hot water heater leaked. That was also three years ago. If only we had known, we could have saved ourselves a lot of trouble and worry.

When did you move in, remodel, get the cat, discover the flooded basement, replace the furnace, install the evaporative cooler, install new carpet, refinish the hardwood floors, replace the attic insulation, etc.?

When did you start noticing the symptoms — maybe not directly, but your first uneasy feeling? Some people don't get "sick" immediately when they move into a new house, but they distinctly remember that they never did like the musty smell in the basement or that the new carpet odor gave them a headache. Even if you can't identify any specific event, you usually are aware that you "just never felt quite right" since moving into the house.

As you read the next chapter on "stealth" sources, keep in mind the question of timing.

- Assume the house *is* the cause.
- Change the interactions of sources with each other and with the occupants.
- Consider what you *haven't* done.
- Examine the timing of the complaint relative to the exposure.

CHAPTER TWELVE

STEALTH SOURCES

Stealth Sources

Understanding common sources and actually finding them in a home is usually very simple. However, if your early attempts have been unsuccessful, then the sources are not so obvious. These are the sources that, like the stealth bomber, aren't easily observed on regular radar screens. These are the sources that our common maps don't include. What makes this task of finding and removing sources even more difficult is the fact that every combination of home and source interaction is different. We need to establish a new starting point for working with difficult sources.

We all know what a home "looks" like. At its most basic level, it has floors, walls and a roof. In reality, there are hundreds of variations; even standard house models from a single builder are not identical. Some differences may be of minor importance, but others may be critical.

For example, is your home a:

- Ranch
- tri-level
- two-story
- mansion
- basement walkout
- garden level
- penthouse
- townhouse
- condominium
- apartment
- Cape Cod
- bungalow
- dorm room

Is the structure:

- Wood frame • brick • cinder block • stucco
  - straw bale construction • metal • plastic • adobe

What other structures and systems do you have, such as:

- Type of garage • type of heating • type of cooling
  - type of flooring • ceiling fans
  - subsurface structures such as a basement or crawl space
  - slab-on-grade • attic • humidification • deodorization
  - filtration • central vacuum • sun room
  - green house • indoor hot tub • exercise room • workshop
  - storage • home office • art studio

Do you use your space to:

- Sleep • eat • watch TV • listen to music • read • exercise
  - play • entertain • work on projects such as woodworking
  - metalworking • photography • oil painting
  - furniture refinishing • leaded glass • collecting • cooking
  - laser printing • laminating plastic picture frames, etc.

What is your life-style or housekeeping habits? Do you vacuum or dust once a week or once a year? Do you have pets such as cats, dogs, cockatiels, iguanas, hamsters or fish? Do you live near toxic factories, smelly feedlots, or in the middle of a virgin forest? Is your landscaping dry, rocky, or swampy, or is your house on stilts over water? What is the age of your house? How long have you lived there? Have you remodeled recently? What is the history or the life-style of the previous occupants?

All of this information is important and any single factor can have a grave impact upon the healthfulness of your indoor habitat. However, presenting a specific analysis for even a few of these categories is impossible. No matter how comprehensive the description, it still wouldn't describe *your* house. There are so many different variables and combinations, how can we possibly describe them in a simpler, more useful way?

Just as we did with sources in general, finding specific sources in a particular house can be greatly simplified. To do that, we have to shift our point of view slightly and give up some old assumptions and myths. Instead of trying to describe and analyze the physical structure or the physical function of the house, as we just did, we need to look at houses from the point of view of *sources*. We start with the three types of sources:

- Particles
- Chemicals
- Living organisms

When seeking the presence or absence of sources in your home, keep these *categories of sources* in mind as you look for:

- **Intrusion** of sources from outside to inside the house.
- **Location** of sources inside the house.
- **Accumulation** of sources.
- **Movement** of sources, or how they get to where you are.

These considerations apply to any type of structure. We can ask these questions as we walk through a house, an office, a school, or even a barn, and generate the vital information we need to understand what is happening. That information provides the basis for generating a plan to stop the exposures.

INTRUSION - HOW SOURCES GET INSIDE

Sources can penetrate a structure from the outside, or they can already be inside the house. They also can be transported inside by the occupants and their pets.

Particles: Dust and pollen can intrude through open windows and doors. A whole-house fan can suck in massive amounts of particles. Particles can intrude not only from outdoors but also from outside the inhabited spaces. For example, the attic is not usually inhabited, but it is typically a very dusty place. Crawl spaces, ducting, and storage spaces aren't inhabited and therefore aren't cleaned regularly. Anywhere dust and debris are allowed to accumulate is a possible source for particle intrusion into living areas. Some materials, such as carpets, clothing and insulation, become sources when they "shed" fibers and fragments.

Particles frequently are brought inside on your own clothes and hair. The fur of dogs and cats can be huge carriers of dust, pollen, and mold.

One house I studied dramatically demonstrated this point. The owners had the ducts cleaned by a reputable company four months before my investigation. Because the amount of debris inside the ducts looked as if they hadn't been cleaned for several years, I began looking for sources that could be sucked into the cold air returns. I found a large return vent on a hallway wall just inside the patio door where the two family dogs came in and went out. The vent cover was black with dirt and dog hair. Now, considering that a vent isn't a filtration device and will collect only a minuscule amount of the total debris, imagine how much dirt actually went inside the ducting and was blown throughout the house.

Chemicals: Although chemicals, such as car pollution and even laundry softener from your neighbor's dryer vent, easily can intrude through open doors and windows, the primary sources of chemicals are usually brought inside your home by you. Actually, the chemicals are part of the manufacturing process or are inserted by the manufacturers. They are packaged enticingly and we trust them so we bring them into our habitat. They become an environmental Trojan horse. Following are some common household sources of chemicals:

- Building materials — These are part of the structure and cannot be removed without tearing down the house. Look for manufactured materials such as particle board, chipboard, carpet, carpet pad, adhesives, and insulation batting. Check the MSDS, if available, for contents. New materials less than three years old still may be out gassing enough to bother someone with a PIR of 4. If your PIR is 5, the out gassing may take 7-10 years before you will be free of complaints.

- Cleaning products — Read all labels and the MSDS. Look for the common problem chemicals such as ammonia, petroleum distillates, chlorinated compounds, alcohol complexes and fragrances.

- Personal care products — Read all labels. Usually no MSDS is available for this type of product. Suspected ingredients include alcohol and its derivatives, glycerin and fragrance. Also, don't be fooled by a long list of natural ingredients. Somewhere buried in their midst is usually a non-natural substance. And "natural" doesn't automatically mean "safe" for you.

Living Organisms: Dust mites, cockroaches and other living organisms can enter from outside a house or be carried in with furniture or appliances. Mold can enter through open windows and doors.

The most common intrusion is water, which generates mold. Water can generate both visible mold colonies on surfaces and subvisible, dispersed mold in the air. Mold inside walls, ceilings and other structures also can generate chemicals called Volatile Organic Compounds, or VOCs.

As you look for water intrusion locations, keep in mind that repeated leaks can be as significant as a single big one. Wet/dry cycles seem to elevate airborne mold to higher levels and at a much faster rate than constant wet conditions.

Notice not only where the water comes in, but where it goes and how long it stays. If materials stay wet 2-3 days or longer, the opportunity for mold growth increases dramatically. Notice the arrows above. Water is obviously inside the wall and coming out through the knot holes.

Snow and rain water can intrude from the outside through the roof, walls, around windows, doors, and other penetrations, and attic vents. Rain and snow also can accumulate in basement window wells, entering either through the window itself or through the foundation farther down.

Ground water typically intrudes into basements and crawl spaces through foundation walls. If the exterior of the walls has been waterproofed, water still can come though cracks. I have even seen houses where water would splash onto the cinder block foundation *above* the waterproofing. Over time, the dampness would creep down through the layers of cinder block, creating visible patterns of white "dust" on the wall. These are minerals left behind as the water evaporates. They usually are not mold, but that sign of dampness is a strong indication of the potential for mold elsewhere.

This basement flooded from a broken water pipe. Water stood 6" deep for several days. The visible mold is colonized mold, which will thrive under certain conditions. The conditions were best between the bookcase and the baseboard. They weren't as ideal beneath the bookcase or above the baseboard.

Ground water also can come up through the basement floor. Even if the floor is never wet, the added humidity often can be enough to promote microbial growth. This is especially true if the basement has been finished and carpeted. To find out, lift a metal filing cabinet and check for rust. I have seen houses where the paint on the basement walls hasn't chipped or flaked in more than 10 years, but floor tile won't stick to the floor and carpet always feels cool and almost damp. In fact, I often find that the life of the carpet on below-grade floors usually is half that on other surfaces.

Condensation in attics and subsurface crawl spaces also can be a serious problem. Moisture in the outside air condenses on cooler surfaces.

Water will actually "wick" up inside a wall, producing structural damage that is visible. But also be aware that the moisture that caused this visible damage is also most likely growing mold and other microorganisms inside the wall.

LOCATION - WHERE SOURCES ARE

Where are the irritants and allergens located? What are the places that you don't commonly think of — forced air ducts, on the tops of door frames, on the walls themselves, under the bathtub, in sealed containers, inside the walls, the crawl space or attic? Are there secondary locations in addition to the primary one? Is the location connected to one source reacting with another?

Particles: Most particles tend to be heavier than air and will settle on surfaces. Look for places you don't routinely clean or never thought of cleaning. Also, very light particles can stay suspended in the air, traveling with normal air currents and collecting on walls, ceilings, drapes and ledges. Forced air ducts are a tremendous repository for particles. And don't forget the carpets. Just because the surface looks clean doesn't mean it's clean below. Furthermore, old carpet generates its own "dust" as it disintegrates and sheds fibers. The most commonly overlooked source of particles is the vacuum cleaner. Not only does it pick up dirt particles from the carpet and blow some back into the air but the bag itself is a "dirt pile" subject to moving air. Also, mold and bacteria collected by the bag can reproduce and be blown out. Bedding and upholstered furniture also are great habitats for pet dander and dust mites.

Sometimes sources are right under our nose. Or feet. Or bed. Other times they aren't so obvious, like where the dust and dander travels on air currents and where it can then re-accumulate.

Chemicals and Odors: Look for source locations of chemicals according to these categories:

- Storage of chemical containers inside the house.
- Opening and emptying of chemical containers inside the house.

- Out gassing of chemicals from "normal" materials.
- Generation of chemicals by living organisms.
- Absorption of chemicals through your skin, especially from cleaning supplies and personal care products.
- Ingestion with food. Although this book does not address food as a source of complaint, it is important to know that chemicals used to preserve food and enhance flavor may cause problems.

Where chemicals accumulate is tricky. Those lighter than air tend to be stronger at the ceilings and on the upper levels of the house. Chemicals that are heavier than air tend to concentrate toward the floor and downstairs.

Pockets of odor and sources in the carpet can be especially tricky. What we smell in these situations usually isn't the original source, but a mixture. The location of the odor may be some distance from where it is noticed. For example, a carpet may have no odor when smelled directly, but the odor-causing complaint doesn't stop until the carpet is removed. Sometimes the source is the carpet pad. Again, you can smell the pad, but the odor it emits is not the same as the one you are complaining about. It is my experience that such odors, especially in small pockets, seldom are themselves the source of the complaint. Finding these multiple sources can be quite difficult, requiring creativity and experimentation.

Living Organisms: Cockroaches tend to be where there is food, such as the kitchen and where kids and pets eat. Dust mites will be in bedding, mattresses, upholstered furniture and carpets. Although dust mites also can be in air ducts, that is not typically a major problem. The mite feces are too heavy to stay airborne more than a foot or so from the vent. However, pay attention to vents right under or beside a bed. Some houses have the vents in the ceiling, blowing downward onto the bed. This could be a prime source of dust mite allergen.

ACCUMULATION - HOW SOURCES INCREASE

Do sources just pile up like "dust bunnies" under the bed or are they alive and breeding?

Particles: Particles can accumulate at the location of the source, for example, dog hair where the dog sleeps. They also can collect in corners and odd places because of airflow patterns. Have you ever noticed how tree leaves in autumn blow with the wind and collect in the oddest places?

Particles indoors behave in a similar manner. We aren't aware of it because they are subvisible, too small to see. They also can collect on walls, ceilings, and other non-horizontal surfaces because of the slight difference in electrical charge between airborne particles and those surfaces. Forced-air systems can collect particles; grow particles, such as mold and bacteria; attract them from unusual locations; and deposit them to even more unusual places.

Who really is in control? You or your exposure sources?

Chemicals and Odors: Usually these are strongest at the source and dissipate with ventilation; however, that is not always true.

Building materials typically become less of a source of chemicals and odors over time. Think of them as containers permeated with a fixed amount of chemical. As they out gas, less chemicals remain inside. However, the more energy-efficient the building, the higher the potential for the chemicals to accumulate. Remodeling brings in new sources while very old materials create problems as they disintegrate, which is a microbial process. Also, living microbes often generate chemicals and odors.

Cleaning products are a common source of chemicals. Think of all the detergents you use for washing, pre-spotting and softening the laundry. As you use the same product over and over with the same clothing and bedding, the fragrances and other ingredients accumulate in the clothing and bedding and inside the washer and dryer. Now add to the mix the dishwasher detergent, floor cleaners, glass cleaners and oven cleaners. Don't forget the really strong solutions for cleaning and disinfecting bathroom fixtures, shower stalls and toilet bowls. Then there are floor waxes, carpet deodorizers, potpourri, plug-in deodorizers, garden and plant fertilizers, and pesticides. These products are usually stored in one or two locations and their chemicals may be mixing and creating who knows what. Even though the containers are closed, no container is absolutely air tight. If your PIR is less than 4 you may have no difficulties. But if your PIR is above 4, then even a few molecules occasionally escaping, mixing and accumulating may be enough to trigger a reaction.

Personal care products are often a "hidden" source of chemical exposures. Few people consider body soap, shampoo, hair conditioner, hand lotion, perfume and other personal care products "chemicals." Chemicals usually are thought of as harsh liquids that smell bad and hurt people. How can your favorite hair spray be a villain? Or that enticing cologne? The key here is that many personal care products are absorbed into and through the skin. The oily ones especially can accumulate in the body. And the slight amounts that escape the containers can mix with each other and with any cleaning supplies that happen to be stored with them.

Mold is the greatly overlooked source of chemical exposures. If the mold is actively reproducing, the quantity of VOCs also is increasing.

Living Organisms: Living organisms breed and travel, then breed some more. If environmental conditions support their survival, they increase to even higher levels. These increased levels can occur in one location or in many. In fact, the secondary source levels often are higher than the original one.

The original source of cockroaches may be a used appliance you just bought. Once inside your house, the cockroaches can find an even better habitat under the kitchen counter by the wastebasket or behind the refrigerator.

Dust mites proliferate where there is an abundance of human skin. They especially love upholstered furniture and bedding. What starts as a small group can easily become huge.

Mold presents similar problems, but with an added dimension. For example, the original source may be a roof leak, but as the mold travels on normal air currents, it can begin proliferating wherever the conditions allow. These places include shower stalls, evaporative coolers, forced-air ducts, humidifiers, crawl spaces and other water leaks. The most important secondary source of mold is the air itself. Once established in the air, mold is difficult to remove. It is not uncommon to remove and replace all water-damaged material in a house yet still have symptoms, because the airborne mold has not been removed. I have seen elevated airborne mold persist for as long a 8 years after the original sources were removed.

MOVEMENT - HOW SOURCES GET TO WHERE YOU ARE

Are you being exposed to those sources by direct contact, like sleeping in a bed full of dust mites or sniffing a tube of glue? If so, you are the one going to where the source is located.

If you are being exposed to airborne sources as they circulate with air currents, the sources are coming to you. With few exceptions, these "breathing zone" exposures are the most likely means of exposure. To appreciate this, consider:

- Air is everywhere.
- Air moves.
- Anything that is airborne can move to wherever the air moves.
- Therefore, airborne sources can be everywhere.

Furthermore, because interior walls and doors are not airtight, air moves throughout the whole house. It moves from room to room, from the crawl space or basement to the attic, under the doors and around the edges of walls. Air moves through any penetrations in the walls and floors that are intended for pipes, electrical wiring, air ducts and plumbing. It may take a day or two, but whatever is airborne will tend to distribute itself throughout the whole house.

Our common-sense understanding of air movement is distorted by our perception that if *we* can't get through an opening, then nothing else can. For example, if we want to isolate the bedroom from the rest of the house, we close the door. While that will block us, the closed door doesn't stop the air from moving under the door or through the sides and top.

This subjectively, egocentric perception of how things work is one of the more overlooked factors in how someone is being exposed indoors. An extreme example is a client of mine who claimed that mothballs could not be a source of exposure because they were still in the box. Ever been near a closed box of mothballs and not smell the odor? It goes right through the cardboard. In one house, the mothballs had been spread throughout a crawl space. When the occupants realized that the fumes were migrating upstairs, they removed them with a vacuum cleaner. First, the vacuum exhaust spread the fumes even more. Second, they left the mothballs inside the vacuum, which they stored in the *bedroom* closet. They assumed that because the moth balls were inside the vacuum cleaner they were isolated from their breathing zone.

Ridiculous? Maybe, but stories like these two are very common.

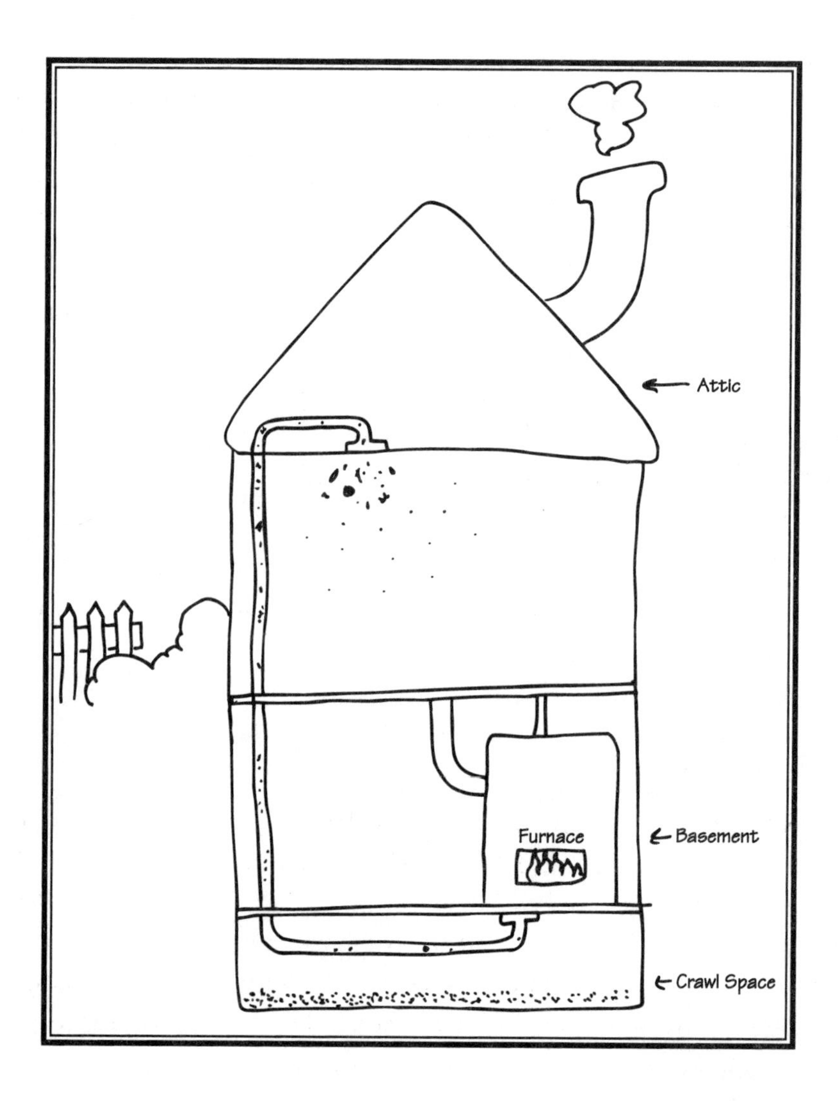

Baseboard heaters, while causing fewer problems than forced air systems, still move the air. As the baseboard units heat the air, they create convection currents. Whatever is in the air will circulate. The accumulated dust, dander and mold will be disturbed and will circulate throughout the house — and they accumulate on the heater fins and on the housing surfaces, ready to be disturbed again. Even the paint fumes from the heater housing may be activated.

If the house has a forced-air heating system or air-conditioning, the movement is magnified hundreds of times. Even with the blower off, the ducts form circulation paths from top to bottom of the house. Add an outside vent for combustion air and you have outside air moving throughout the house. An example of this is a house that had a musty smell in the bedroom vents early in the morning. The outside vent was two inches above old wood chips on the ground. When the cool outside air moved into the basement, it carried the odor of deteriorating wood with it. As the air moved through the ducts from basement to bedroom, the odor went with it. The solution was to clean up the area directly below the intake vent.

Do you realize there are openings in the ducts themselves? Forced-air ducts are not designed to be airtight. All corners, seams, joints and connections allow air to move back and forth through them, permitting particles to enter the interior of the ducts and accumulate there. And once inside the ducting, those particles can move with the air currents and travel to wherever the air goes.

When the furnace blower actually is running, the air and all it contains is forced throughout the house. Consider that smelly basement, attic or crawl space that have forced-air ducts running through them. That is how the air in those locations gets to where you are. As far as exposure sources are concerned, you may as well be living right in your crawl space, basement, attic, garage, workshop or chemical storage area.

Situations range from the most complex and disruptive to the simplest. Use your Personal Impact Rating (PIR) as your guide for the level of diligence required for your situation. The higher your rating, the more detailed you should be. The lower your PIR, the more general and superficial you can be. Do only the minimum necessary to stop your personal complaint. You do not, and in fact should not, take all the actions specified here. You should only do what is necessary for *your* circumstances.

Chapter Thirteen

Stealth Impact

Stealth Impact

Not all difficulties with stopping complaints have to do with physically removing sources to below your level of susceptibility. Other factors involved can be quite confusing when trying to connect susceptibility with exposure. Some helpful descriptions and analogies are:

Sensitization - With continued exposure to non-complaint levels of a particular substance, you may gradually become sensitized to it. That means even though you never previously reacted to it, you do now. Sometimes this reaction also becomes generalized. You now react to *other* exposures that were previously not a problem. A common example is reaction to a long-term, low-level natural gas leak. The gas odor is only occasionally noticed and then seems to be gone. But the gas is still present. Over time the more susceptible individuals may begin to react to the gas plus a variety of other substances.

Total load - All the exposures combined may be more important than any single exposure. While each specific exposure to, say, cat dander or mold or perfume may not cause a reaction, the combination of them might. Another way of saying this is to ask if you are looking at the forest or the trees. The fact that you don't crash into one specific tree doesn't mean the forest itself doesn't exist. You could still hit anyone of them at anytime. A clue to total load is if you become more reactive to certain foods, dander, odors, etc., at times of increased susceptibility, such as when you are ill, stressed or overly tired.

An overloaded body won't be as visible as this overloaded truck. Nor will it be contaminated with such toxic substances and still be alive. But long-term exposures to low levels of toxic and nontoxic substances alike can often influence a sequence of ill-health events.

Triggers - This is best described as "the straw that breaks the camel's back." While the specific "straw" that triggered the breaking of the camel's back is important, it may not be the primary problem. The total load of all the big bales of straw that the camel was carrying is most important. The camel can carry many bales, but at some point he can't carry any more. At some point just one more straw will be too much. That doesn't mean *that* exposure is the critical one. It wouldn't have triggered anything if the total load were smaller.

The straw that breaks the camel's back typically isn't as important as the total load. The trigger could have been anything.

What is now important is that the damage has been done and the burden must be greatly reduced before the camel can heal. Removal of only the trigger will be insufficient.

Look at all the components of the total load and prioritize your actions based on your PIR.

When a breakdown occurs, removing just the straw that triggered the problem will not be sufficient to stop the suffering. The rest of the load will have to be removed. The camel cannot heal as long as he is burdened with that load. A clue to triggers is when removal of a specific source doesn't result in a rapid removal of the symptoms. If the symptoms return or tend to continue for a prolonged period of time, you may need to remove additional sources.

Masking - Masking is one of the more difficult concepts to understand. Damage is being done, but there are no perceived symptoms. Or at least there are no negative symptoms. An example of masking is when extremely reactive people claim that smoking cigarettes makes them feel better. It has been theorized by some that the intake of the smoke, nicotine and tars "loads" the body sufficiently to camouflage the pain and discomfort to other exposures allowing them to go relatively unnoticed and thus feel more comfortable.

Another example is the use of air deodorizers. They may be bottles, plastic discs you stick on the wall, baskets of potpourri, or ones you plug into the electrical outlet. None removes an offending smell. They just cover it up. Now that's no problem if an unpleasant odor is the complaint. But, if you are reacting in some way, deodorizers just confuse the issue. In addition, you may be susceptible to the deodorizer ingredients.

Delayed Reactions - Not all reactions occur immediately upon exposure. Sometimes they can be delayed by minutes or even hours. This can make it very difficult to identify sources, especially if several exposures occur before the reaction sets in. For example, many susceptible people experience a delay of as much as eight hours after exposure to mold or other substance. In the meantime, they have been in several houses, including their own, driven a car, been to the doctor's office, and eaten at least one meal. They usually blame the most recent exposure that they are aware of as the cause of their symptoms. It takes some pretty motivated detective work to successfully differentiate sources if the complaint is delayed.

Mis-association - Not all reactions are to what is most visible or otherwise obvious. A classic example is people who suffer from tree pollen in the spring. They point to all the "cotton" flying around from cottonwood trees and comment about their cottonwood allergies. It sounds OK, and most people — including many allergists — don't dispute it. The suffering is unmistakable and the source seems obvious. However, the "cotton" is not the pollen. The tree pollen was present two to three weeks previously. The cotton is the *result* of the pollination. The current suffering is to a different source. Mis-association of indoor sources most often occurs with odors. What you smell is what you blame. You may be right, but frequently the real source won't have the most noticeable odor or any odor at all. You will need to investigate further.

Another type of mis-association occurs when authorities attribute sufficiently individual experiences to illusion or hysteria. The patient, if they place their trust in that authority, starts down a new road that does not lead to healing and health — but only to blame and victimization.

Another form of mis-association occurs when the patient is prescribed a regimen that they don't fully understand or trust. They can neither successfully complete the plan nor evaluate the results. An extreme example is for you to follow the treatment plan of a primitive witch doctor. Or for the witch doctor to follow a modern, technologically advanced protocol.

The Weakest Link - Pull on a chain long and hard enough and it eventually will break. That can be predicted. But what cannot be predicted is which link will break first. What is your weakest link when your body is under attack? If your PIR is low, your reaction may be a runny nose or watery eyes. A moderate PIR may result in a headache, stiff neck, palpitating heart, sore joints or muscles, or a rash on your arms, legs, stomach or back. A high PIR may be experienced as chronic fatigue, migraine headaches, sleep disturbance, depression or hyperactivity. Everybody has a different set of susceptibilities due to their genetics and previous exposures. The confusion usually arises when you hear yourself saying, "But it can't be *that* because..."

As is obvious by now, all these complicating factors can make virtually any link between susceptibility and exposure as invisible as a "stealth" bomber. The most effective response is a one-step-at-a-time approach with lots of patience. Otherwise you'll lose track of which influences what.

Chapter Fourteen

Walking Through a House

Walking Through a House

All the previous chapters have focused on indoor exposures from a relative distance. The descriptions have been about "those substances out there" that just happen to be inside many houses. But when the focus shifts to you actually being inside the house, being directly exposed to all these substances, your experience suddenly becomes more relevant and personal. The starting point of this chapter will be a personal experience of a house from the inside. It explores the features, systems, activities and exposure sources where people actually live — or where those sources are living with you often like a rude, obnoxious houseguest.

As we walk through a typical house in this chapter, imagine that it is your house. Focus on the rooms of your house and your activities therein. Pay attention to key features of the house, such as the highest point and the lowest point, dark rooms and sunny rooms. Notice, too, key systems of the house, such as those for heating and cooling, and how they affect air circulation and moisture content. Finally, pay attention to actual air circulation patterns and anything that moves air or influences where air comes from and where it goes. Consider how it gets into the air you breathe, your "breathing zone."

As you walk through this hypothetical house, **consider *everything*!** Remember, what you thought you knew and what you have already done hasn't worked. The solution to solving your particular problem may be much different from what you anticipate or assume. Check everything. Write down all possibilities. You can sort it out later as you develop and refine your plan in the next chapter.

Let's now walk though a typical house together with a new perspective on how we organize our information and experiences.

THE HIGHEST POINT

Begin the tour of your house at its highest point and work your way down. It makes little difference if the highest point is an attic, top floor or ceiling. Source intrusion from beyond your living space, such as the apartment above you, can be brought into your plan later.

There are several reasons for starting at the highest point. Lighter-than-air substances will accumulate at the top of rooms and buildings. The roof is the most likely source of water intrusion. Finally, because most substances are heavier than air, the upper level is the least likely to contain exposures that will overwhelm your susceptibility and mask further reaction.

THE LOWEST POINT

The lowest point is often the basement. This is where most of the central utility systems are located, such as the forced-air heater and air conditioner or the boiler for baseboard heat. Consider how the utility system can generate or affect airflow patterns. If the lowest point is below grade, is the floor composed of bare dirt, a cement slab, or a wooden structure over a crawl space? The lowest point is also the most common location of the most difficult and overwhelming sources.

GROUND LEVEL

This is the most common level in houses. It is important to make note of what is above and what is below this level. Is there another story above or just the roof. Is there an attic? Is the floor slab-on-grade or is there a basement and crawl space below?

BEDROOMS

Length of exposures are often longest in the bedrooms, where you spend the majority of your time in the house. Are your symptoms better or worse when you get up? Some occupants notice symptoms immediately upon getting into bed, yet others wake up in the middle of the night. Have you replaced the mattress in the last few years? Have you replaced any of the bedding recently? Are the mattress and bedding very old? Did your symptoms increase or decrease within a few days of when the changes were made?

The bedroom also is where you are most likely to come into direct contact with many potential sources, including dust mites. Chemicals from new bedding, mattresses, bed frame and waterbed bag are common. Other chemical sources are the detergent and softener products used on the bedding and on the clothes in the closet. Mold typically will be found in very old material. Pollen, which collects on hair, is transferred to the pillow.

Don't forget your favorite down comforter! In a number of houses where the occupants were suffering, the only problem was the down bedding. Sleep without it for one night. Then you will know whether it is a cause or not.

KITCHEN

Common chemical sources in the kitchen are the gas range and oven, cleaning products stored in the pantry or under the sink, and new building materials. Mold often is unseen, but common nonetheless, under the kitchen counters and dishwasher, behind and beneath the refrigerator in the drip pan of frost-free units. These are also typical areas to find cockroaches.

LAUNDRY ROOM

Chemical city! The laundry room is where the detergents and fabric softeners are used. It is also a common storage area for floor wax, silver polish, shoe polish, pesticides, plant fertilizer and a cornucopia of detergents and other cleaning products. Check also for leaks from the washing machine and hoses that can foster mold growth.

Be sure the dryer is vented to the *outside*. (By outside, I mean outside of the *house*, not just outside of the laundry room itself.) Dryers vented into attics, basements, and especially crawl spaces are trouble in the making. Dryers blow warm, moist air full of lint into a dark space. These are ideal conditions for growing mold, bacteria and other living organisms.

GARAGE

Do you frequently smell garage or car odors inside the house? Garages are common storage areas for even more chemicals than laundry rooms. Garages also hold paint thinner, paint, gasoline and oil. They are great storage areas for old wood, cardboard boxes and other food for mold. And mold, as well as garage odors, can circulate into your living area through forced-air ducts, floor and wall penetrations for wiring and plumbing, and normal use of the entrance door.

PETS

Where are the pets allowed to roam? Where do they sleep? Do they frequently go outside and then back inside? How many and what kinds of pets do you have? Have you been tested to determine whether you are allergic to any of them?

CRAWL SPACES

If you live in an area that is too damp or too rocky to have a sub-grade crawl space, count your blessings. In my opinion, crawl spaces should be outlawed. In more than 12 years of inspecting and investigating houses, in only five have crawl spaces *not* been part of the problem.

How long has it been since you even looked inside your crawl space? Go look now. Notice the open dirt, the water marks on the foundation walls, the white powder on the edges of the dirt clods, the spider webs, the mouse droppings, the musty odor. Is the air foul-smelling?

The "whitish" areas are efflouressence. The material is minerals left behind as water vapor slowly penetrates and then evaporates. It is rarely mold.

You wouldn't allow any of this in your actual living areas. So why allow it here? A crawl space is not unlike a back corner of a forgotten closet. Whatever is in the air of the crawl space is in the air of the rest of the house, especially if you have a forced-air heating or cooling system with ducts running through it.

But crawl spaces can be made even worse by using them for storage. If you store cardboard boxes of clothing, papers and books in them you are providing even more food sources for mold and bacteria.

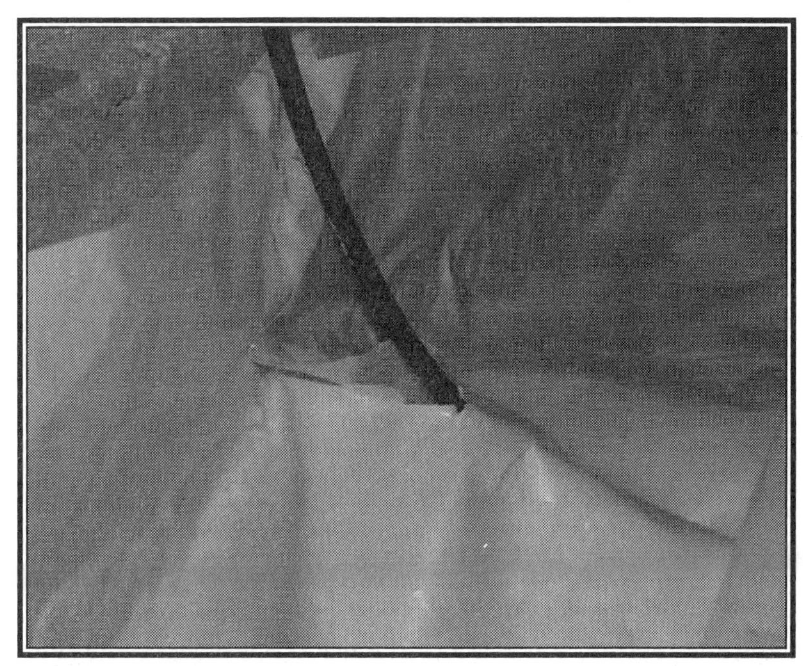

Penetrations thru the plastic, as shown here with the slit for the cable destroys the integrity of the vapor barrier.

If the dirt is covered with plastic, are all the seams and the perimeter sealed? Holes and tears also destroy the integrity of the vapor barrier. Are there any vents to the outside? Are the vents open or closed?

BASEMENTS

Next to crawl spaces, basements are the next most likely location for sources of exposure, for many of the same reasons. One difficult situation to identify and mitigate is when the basement walls are finished with drywall or paneling. Moisture — not necessarily running water, but just humidity over 70 percent — can, over time, generate particularly offensive and dangerous microorganisms. Some of the organisms, other than mold, that have been identified include Stachybotrys bacteria. Under extreme and chronic conditions Stachybotrys has been implicated in damage to the immune system. You won't know it's there until you open the walls and look. And you will be reluctant to take such extreme and disruptive measures unless you have good reasons as to its necessity.

HEATING AND COOLING SYSTEMS

With forced-air, look for small openings in the ducts at corners, joints, seams and connections. These seemingly insignificant holes can suck in contaminated air and blow it throughout the house. If there is anything in the air that you don't want to breathe, the ducts could be a factor you want to address. However, if the air is better there than in other parts of the house, don't close the holes. Cut a bigger one! Let more of the good air circulate to the habitable areas of the house. We aren't interested in *the proper solution* from a building-trades point of view. We want whatever will successfully remove, isolate, or reduce the sources of our exposures.

This furnace is installed on dirt in a crawl space. A recent flood left muddy water marks on the side of the return duct, meaning dirty water was inside the air flow system. This must be thoroughly cleaned and disinfected.

Other problems associated with forced-air systems is insulation inside the ducts, extra-high-efficiency units that require a heat transfer liquid, mold from humidifiers, and mold from the water drip-pan under the air conditioning cooling coils.

Central humidifiers, especially the old reservoir types, are often a breeding ground for mold and other microorganisms. Some squirt a mist of water into the air, which leaves deposits of minerals that can flake and shed particles that blow throughout the house. If you must have a humidifier, install the newer flow-through type. But make sure they use a metal grid and not a spongy pad.

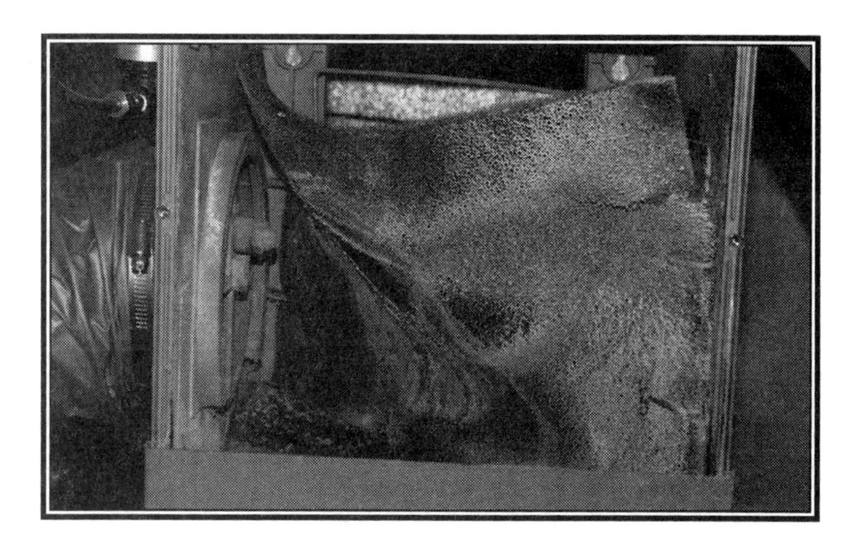

This humidifier is a flow-thru type, but with a spongy pad instead of a metal grid. The minerals, mold and other potential contamination is obvious.

EVAPORATIVE COOLERS

As with crawl spaces, these, too, should be outlawed, in my opinion. In 10 years, rarely have I seen an evaporative cooler that was not a major part of the problem. I have heard of only two people who were susceptible to mold and claimed to benefit from evaporative coolers over the long term.

Not only can they be a huge breeding ground for mold and other microbes, the increased humidity in the house often supports elevated levels of those same microbes.

ATTICS

Problems with attics are rare and almost always involve repetitive roof leaks. However, when attics are a source of exposure the solution can be expensive. The typical problem is insulation that becomes a breeding ground for mold and other microorganisms — a problem amplified by further water leaks. Once the insulation becomes contaminated, it has to be removed, the attic area decontaminated, and new insulation installed. This then increases the risk for people with a high PIR for chemicals.

SUN ROOMS, GREENHOUSES AND HOT TUBS

As wonderful as these are to have in a house, if there is airborne mold, these usually contribute to the problem. They may not be the original source, but the warmth and humidity typically elevate ambient mold levels. Also, the chlorine from hot tubs can be a respiratory irritant. (And no, the chlorine does not kill off the mold in the air.)

In summary...

Evaluating a house for sources of exposure to allergens, respiratory irritants, toxins, or triggers for asthma, fatigue and headaches, often requires more investigation than merely locating the most obvious source. If you have had difficulty eliminating your complaints, it is important to understand the characteristics of the particles, chemicals and living organisms that are in your house. In addition, you also must solve the complex puzzle of the interactions between the sources; the sources with the environment; and both of those with the occupants.

The greater the impact of those exposures (the higher your **PIR**), the more difficulty you will have in identifying and reducing your exposure complaints. As this process unfolds and you discover more complexity, the impact of the dilemmas identified in Chapter 9 may intensify.

This exponential increase in complexity makes what seems like a simple goal extremely difficult and frustrating. The techniques, although sometimes confusing, are relatively simple. Putting it all together for *yourself* is what takes the extra effort.

If this has been your experience, now may be an opportune time to review the minutia in Chapter 8, *Sources and Their Removal*, and Chapter 10, *Evaluating Products and Services.* Although those chapters may have appeared, on first reading, to contain more detail than you would ever need to know, now is when one of those minor details may provide the key to successfully completing your journey. As you increase the detail and relevance of the information generated by your feedback loop, your struggle will begin to gradually lessen and your chances for success will increase.

Chapter Fifteen

Your Personal Plan

Your Personal Plan

It is now time to develop and execute a personalized complaint removal plan for *your* house. The process is a repeating feedback loop designed to generate new information that you can trust, when you need it, for making decisions about what to do, what to avoid, and how to evaluate the results. Expect to execute the plan several times, with each successive attempt modified by the new information generated from previous ones.

YOUR GOAL

Step 1: Why do you want to clean up your indoor air? What complaints do you want to remove?

YOUR STARTING POINT

Step 2: Write down everything you are aware of being susceptible to. Include not only what an allergist has tested, but anything else you have complaints about. Perhaps there are some things, such as perfume or laundry detergents, that you don't associate with a complaint or a problem but you just don't like. They may not be a primary cause but part of the total load.

Step 3: Write down all the symptoms and complaints you can think of.

Step 4: Associate, as specifically as you can, each symptom with a suspected exposure or combination of exposures.

Step 5: What is your best estimate as to the cause of your complaints?

Step 6: Review the history of your house. How old is the house? How long have you lived there? What information do you have about previous occupants? Did they, or you, have pets,

Notes

smoke cigarettes, remodel, or neglect to repair roof leaks, plumbing leaks or basement floods? Think of anything that could create, cause, or support a source of exposure.

Step 7: Compare the *timing* of your symptoms with events in your house.

EXPLORE THE TERRITORY AND EXPERIENCE YOUR HABITAT

Step 8: Walk through your house as described in the previous chapter. Notice anything and everything that could possibly be a factor in your complaint.

Step 9: Observe your symptoms, especially the subtle ones, as you go through your house. If they change as you move from one location to another, remember the associations from Step 4. Can you find the sources associated with the symptoms? Are there symptoms that surprise you, such as chemical reactions, but you don't see any chemicals? Perhaps the chemicals come from cleaning products under the sink or from personal care products in the shower stall or from the bedside table. Pay particular attention to any detectable smells and odors, whether pleasant or foul. If you cannot detect any differences anywhere, that is a strong clue that you may have a "smoke-filled room" phenomenon that is masking your reactions. Something is pervasive and may have become "normal" to you.

An example of this is a client whose house smelled so moldy that just opening the front door made me back away. Though she was quite ill, she claimed not to notice any odor. Once the source was removed, she felt much better and became quite healthy. But she wasn't happy with her house. She said it didn't smell like home anymore. So beware of your preferences!

As you continue through your house you may encounter several exposures that cause concern. Instead of assuming that whatever you detect is harmful, stop and ask yourself the question, ***"Is this really harming me, or is it merely something I happen to notice?"*** Give yourself some time to evaluate your experience. If you discern that it is harmful, then make that specific item a high priority with your plan. If it is merely unpleasant then make it a lower priority.

Notes

"DRAW" YOUR MAP

Step 10: Sit back down and establish your PIR, overall and with each specific exposure or combination of exposures categories.

Step 11: Establish your PIR for each area of your house. Where were you reactive? Where were you *not* reactive? Do you naturally gravitate to some locations and just never spend any time in others? Why?

ESTABLISH A NEW STARTING POINT

Step 12: What do you now think is the most likely cause or causes of your highest PIR complaint? Which type of source: Particles, Chemicals, or Living Organisms?

Step 13: Determine how best to stop the exposure:

- Remove the source.
- Isolate the source.
- Dilute the source by ventilation.
- Reduce the source by filtration
- A combination of the above.

Step 14: Establish your specific plan as to:

- What to do — based on your PIR.
- The sequence, so that one step doesn't recontaminate a previous one; e.g., clean the furnace ducts *after* the hardwood floors are sanded and the new drywall is installed.
- What to avoid — so you don't replace one problem with another.
- How to evaluate the results. Will success be determined by the removal or reduction of symptoms, a lab test result, a behavioral change in your children or your pet?

Step 15: Locate and evaluate the necessary products and/or services. Establish the costs involved.

ESTABLISH BOUNDARIES AND LIMITS

Step 16: How much time, effort, and money are you willing and able to commit? Is it sufficient to execute the cleanup? If

Notes

not, how much can you do? What actions will get you the most results for the least cost? What is the possibility that spending money on the cleanup will reduce costs for medical bills and prescriptions? What is the balance of benefit vs. risk? How can you change your plan to stay within your budget? At what point do you decide to move?

EXECUTE THE CLEANUP AND THE CHANGES

Step 17: Set the standards for any work to be done. If this work is done by others, state clearly — in writing — any guidelines for what to do and what to avoid. Don't ask for any opinions of safety. You are the only one who can determine that. Manage the process and be personally involved — as best you or a trusted friend can be — in the process, including *constantly inspection* to ensure your instructions are accurately followed. Serve *your* best interest, not anyone else's.

EVALUATE

Step 18: Evaluate the results. If your efforts are successful, congratulations! You are done. Make note of what was most effective and what wasn't. Should you ever have to do this again, this information will be invaluable, saving you lots of time and effort next time. And the experience will be applicable if you look for a new home.

If you were not successful, why not? The most likely reasons are:

- The target source was not actually removed.
- Not enough of it was removed.
- It was removed from its original location, but allowed to recontaminate your habitat by being blown or otherwise deposited elsewhere.
- It was successfully removed but was replaced with any number of other multiple source. Multiple sources should be suspected if your symptoms initially improve but then seem to reoccur. The "smoke filled room" analogy is probably at work. One or more sources were removed but over then next few days to weeks, your house is filling up with more "smoke" from the other sources.

Notes

Evaluation of any plan is the most critical and the most difficult task to successfully accomplish. The results of your evaluation will determine what you do next. The information you generate and then interpret directs you to your next starting point. If your next starting point is incorrect, you most likely will fear that you will become lost again, not knowing for sure where you are, where to go next, or how to get back to the previous starting point.

However, all is not lost. It's frustrating, aggravating and perhaps expensive, but the information you need has actually been generated. The very fact that you were wrong only means that the desired result has not been accomplished. Don't take it personally — even if others do. The lack of success is important information. It is just as important to know what *not* to do as it is to know *what* to do.

If the change was positive but insufficient, you have improved your indoor habitat, which is also critical information. The lack of success is due either to not having removed, isolated or reduced enough of the target substance to stop your reactions; or other reactive agents are present. Perhaps they were there all along but were masked by what you successfully removed. In this case, make your best estimate and start again.

Clues for resetting your starting point will be based primarily on your symptoms. Have they changed? Have they been replaced with something new? Are they gone? Are they better, but still not tolerable? The next step is to review the factual information about the category of exposure you are attempting to stop. Then review your PIR information to guide you in reestablishing your priorities and determining your need for diligence.

REPEAT AS NECESSARY

Step 19: If your results were not sufficient, return to Step 1 and repeat. However, as you clarify your new starting point, also reevaluate your limits and boundaries. Even if you can afford to continue your efforts, other factors may have intervened in the meantime making it more feasible to move to another house. Only the simplest situations will be solved the first time. The others may take two, three or more attempts.

The evaluation step is the most critical part of the process and the one that affords the best chance for healing. It is the step where you decide whether to stop or to continue — and why. It is also the action that completes the feedback loop that is so critical for success.

My most successful clients have been the ones that loop through these steps in fairly rapid order — one step at a time — creating a "tight" feedback loop. In fact, working the feedback loop is, itself, often a healing process.

WHAT ABOUT SHORTCUTS?

Three primary actions usually will give you the biggest immediate improvement:

- Duct-cleaning
- Carpet cleaning
- Wall and ceiling cleaning

So why not just skip all the philosophy, politics, investigation, detail and hassle? Why not just do these three things and get on with your life? If your situation is that simple and you have the money, go ahead; that will be your plan.

However, these three actions mostly address *particle* sources. They remove the *accumulation* of particles but do nothing to solve the problem of chemical sources or living organisms. Neither do they address the *sources* of the particles. In addition, the cost for doing all three together may be prohibitive; one or more actions may not be necessary; and the results may be only temporary if the primary sources aren't first removed or isolated. In that case you will have to do it all over again.

You need to make a personal judgment about what your goal is and where your starting point will be. You may solve your complaints very simply, but may also experience varying degrees of complexity. Start somewhere, evaluate the results, then loop back again until you get the results you desire.

As you continue to "loop" through this cycle of starting point-action-evaluation-starting point, you will become more familiar with your indoor habitat, its occupants, and all the interactions between them. Your process of eliminating possible causes will strengthen.

I was an active, vital, healthy 45-year-old woman. I had advanced well in my career and was thoroughly enjoying life. My kids were nearly out of school and I was looking forward to traveling with my husband.

Then they moved my office into a brand-new building. I started getting headaches and then after a week they became migraines. I hurt all the time and couldn't sleep. My work suffered and I started making dumb mistakes that I hadn't made in years. But worst of all, I started reacting to my own house! I had never noticed a problem before but now it's intolerable.

I've tried talking to my boss and the building manager but they said it was all my imagination. They had used the best contractor in town, checked the MSDS on all materials used and had confirmed that there were no OSHA violations. And, they suggested, since I also felt sick at home then that's where the cause was.

So I cleaned up my house and went away for a week. I was reaction-free when I came home and stayed that way until the first day back at work. Wham! Within 30 minutes the migraines were back. And then I found out that several other people were hurting also. And only when they came to work.

As we talked about our experiences — all the while being accused of being busybody women on the verge of hysteria — we began to realize that only certain parts of the new building were affecting us. Also, as we paid attention to what was happening, we agreed that we constantly smelled diesel fumes. Further investigation found the HVAC fresh air intake was right over the loading dock where big semi trucks sat with their engines running.

We weren't hypochondriacs after all, and the new building materials weren't the cause either. And the remedy was simple! But management would do nothing because no regulations were being violated and we couldn't scientifically prove our point. Do I stay and suffer or do I risk destroying my career but in good health? Why can't they just fix the problem?

Chapter Sixteen

How to Stop Being a Victim

How to Stop Being a Victim

This chapter will not be applicable to the majority of readers. Most of you either will have a PIR of 2 or 3 or will have successfully reduced your exposures and improved your susceptibility so that the effects are no longer intrusive or disabling. However, a few of you still will be victimized.

This is not meant to be a definitive psychological treatise on the abuse and victimization of people with a chronic disability. I leave that to experts such as the three in the final chapter. Rather, this is a starting point to begin a process — and to come back to — when you lose your bearings. Here is a summary of what I, my friends and clients have learned.

My original title for this chapter was "*How to Avoid Being a Victim*." But several people pointed out to me that focusing on "avoidance" could inadvertently lead to denial. What is needed is neither denial nor hypervigilance. Rather, what is most successful is awareness of the dangers to you, the effects of various exposures and how to make choices to stop the exposures. Also, people with a high PIR, by definition, are being victimized. Public health and safety standards, while designed to protect the majority of the population, also exclude others.

However, that does not mean you have to continue acting like a victim. Nor does it mean that you cannot have a meaningful life with self-respect and value. You may have this challenge the rest of your life and your financial and energy resources may also suffer, but you still have choices.

In Chapter 1, *My Starting Points*, I talked about the impact of the video by Ken and Maggie Dominey. To reiterate, I realized that:

1. This illness is real and should be treated very seriously.
2. Something can be done about it.
3. I was not alone.
4. Although I had been victimized, I didn't have to stay a victim.

Although you had no choice about what happened to you, you can choose how you are going to respond to it. Although you may be disabled, you can still choose how you want to spend your life within your limitations. Perhaps all you are capable of doing is talking on the telephone for only a few minutes at a time. But your knowledge and experience may be exactly what a person new to this phenomenon needs. What a loss if you choose not to share your experiences, to not give whatever you are capable of giving.

The most powerful action a victim can take is, make choices.

Then continue to make *more* choices as you:

- Seek appropriate medical help to be sure you are not suffering from a standard medical condition. Attempt to find a doctor who also is familiar with environmental exposures as a source of symptoms.

- Seek alternative help, but only from those who truly understand the role of exposure.

- Seek out others to help clarify and expand your understanding so that your next choice is even better.

- Avoid people who will listen to your endless complaints and then contribute their own endless list. That behavior will only reinforce your sense of victimization.

- Seek psychological help, for the purpose of learning more effective skills for coping with a chronic condition and to strengthen your mind and body. Avoid anyone who blames you for your suffering. Your psychological dynamics didn't cause the illness, and neither will it cure it. But it can interfere with your doing what is necessary to stop the suffering. Likewise, it can support you to take productive action.

- Seek spiritual help, whether from a traditional religious source or in some other form. The majority of sufferers report strong benefits from the knowledge that they are not alone and that there is more to life than their suffering. There is a fundamental difference between the suffering owning you and you owning the suffering.

- Give. Contribute. You may be severely impacted but you are not helpless. You can still make the choice to live as fully as possible during *any* level of impact.

> *Should not the giver be thankful that the receiver received? Is not giving a need? Is not receiving, mercy?*
>
> *— Friedrich Nietzsche*

- When others attack you, try not to respond personally. They are acting out of fear that they may one day be impacted as you are. Keep in mind that you have had a life altering experience — and it has not been a shared experience. Even those with similar experiences may be harmed by the very procedures that save you. Those who have not had a similar experience will have little if any comprehension of what life is like for you and what you have to do just to get through the day. You are talking a foreign language to them, and they to you. Huge misunderstandings and emotional turmoil can result. They can't understand any better than men can understand what it is like for a woman to deliver a baby. And you can never fully explain it to them. Remember, most people "need" us to be their "necessary victims" so they can continue the illusion that they are safe and secure and therefore avoid acknowledging their fear.

Rather than responding directly to their attacks, assume that their behavior is due to a lack of information. Assume they would prefer to understand and support you. Ask them questions like, "Would you like to hear more about this illness?" Or, "There have been several new studies that are not only interesting, but hopeful." Or, "That's what even the experts thought until researchers discovered..."

BUT SOMETIMES A DIRECT APPROACH IS APPROPRIATE

Elsie, one of my clients, relates the following incident with her anesthesiologist. Although her actions in this particular situation cannot be recommended as a standard response, it did work for her. It also dramatically illustrates the importance of being personally and actively involved in your efforts, and asserting your own best interest rather than meekly deferring to authority.

"We've all had the experience of dealing with doctors who think we are only imagining we are sick. Recently when I was waiting for a surgical breast biopsy I encountered an anesthesiologist who was the worst ever. There I was, a crippled, old, gray-haired lady lying on a Gurney with a couple of big wires sticking out of my breast, waiting for surgery. I had asked to see the anesthesiologist and have him read some literature on the special problems that chemically sensitive people such as myself often have with anesthetics. I also had some personal notes about which ones had previously worked for me.

"When he arrived, he didn't even introduce himself. After he asked me what my symptoms were, I recited a partial list of my previous problems with aesthetics that included confusion, disorientation, visual disturbances, violent gastric disturbance, various peripheral neuropathies, arthritis, etc.

"'Those aren't *symptoms*,' he exploded, 'those are ordinary *side effects!*'

"He then told me I shouldn't be reading the kind of things I was reading without also reading proper medical literature. (How did he know what I was reading?)

"Then I exploded, too, and yelled 'Bullshit! If you didn't want to hear about my symptoms why did you ask?'

"Now, I don't recommend yelling at doctors when they have your life in their hands. But the roof didn't fall in. In fact, he sat down and read every page of my folder of information. He said he'd honor my requests.

"After the operation I had no serious side effects — or symptoms — and the surgeon confirmed that my recommendations had been followed.

"Maybe the moral of this story is, don't apologize for being chemically sensitive, and for doing whatever is necessary to get their attention. By the way, if I have surgery there again, I plan on asking the same person if he will **please** be my anesthesiologist."

Elsie's story illustrates several salient points. To prevent a potentially dangerous situation of being administered anesthetics that could cause unacceptable or even life-threatening side effects, she prepared herself with accurate information that she could trust. She then requested a meeting with the anesthesiologist so he would understand her concerns and honor her needs.

NOTE: Notice the different interpretations given to the effects of the anesthetic. What Elsie called "symptoms"–vague clues of possible harm–the doctor called "side effects"–anticipated events that could be safely ignored.

However, the doctor's initial demeanor and lack of cooperation left Elsie in a dilemma. She could have just kept quiet, desperately hoping she would either be safe or that she could tolerate what was going to happen to her. She could also have cheerfully "just trusted the doctor." After all, he is the expert, not she. And he has (hopefully) read all the proper medical literature, including that on chemically sensitive people.

Either response meant giving up her best interest for that of someone else. But why do that? She's the one with the concerns and the needs. She's the one who has the right and the final obligation to make choices that influence the safety, survival, and the enhancement of her life.

Elsie made her choice and challenged the authority. She chose to directly experience her life at that moment. She was fortunate that this anesthesiologist responded professionally and complied with her request. But even if he hadn't, the end result may have been postponing the procedure, thereby preventing a potentially life-threatening event. Identifying and enforcing her best interests was a win-win choice.

Finally, if Elsie were truly "crazy," her actions toward the doctor would not have changed after the successful procedure. If she were merely "crazy," she would not have changed her actions in step with the changes in her "total set of environmental parameters." Because her responses did change, it is more likely that her outraged response was *appropriate* to an outrageous situation.

This example does not give license for *any* behavior. Appropriate behavior can only be determined by open, honest communication between the parties involved that has the best interests of all at heart. Elsie's story underscores a desperate need for the medical, psychological and patient communities to establish appropriate and effective guidelines for behavior.

Joe had a PIR of 4-5 for most categories of exposure and hadn't been able to work for a year. Life was tough for him, but at least he was able to stay relatively complaint free in his home. The house was familiar and he knew exactly where he was safe and where he would react. He could anticipate most situations and knew exactly how to respond.

But life was also getting tough for his wife, who wanted nothing more than to please him and to help him get well. While he was out of town for a couple of weeks at a special medical clinic, she decided to surprise him. Relying on her own experience of what was safe for Joe, and by consulting with the experts at his clinic, she had the house professionally cleaned and several rooms repainted. She knew that was risky but she also knew that Joe could tolerate non-VOC latex paint after four or five days. She converted the spare bedroom into a "safe" room so he would have a non-reactive place where he could heal and rest. The work was done on schedule, according to her instructions. She had a friend who had similar sensitivities check out the work. She even rearranged the living room. Just before he returned home, Joe reported that he felt much improved, and both he and his wife were excited about a successful recovery. Everything was ready for the Joe's return home.

But when Joe walked in the front door he was not pleased. In fact he was irate. "What happened to my house!" he exploded. "Why did you move my clothes? How do you know the painter actually used the right paint? Did you see the paint can? Show it to me so I know!"

What Joe's wife and the clinic had inadvertently done, in the midst of all her love and their good intentions, was to leave him out of the process. His landmarks, his map of the house, were destroyed. He no longer knew how to interpret his reactions. He had no idea whether a change in his body or mood was good or bad, where the exposure came from or how to respond. His old starting points were gone and he would have to reconstruct them. The moral of this story: It is absolutely essential that the person with the complaints be an active participant in ***any and all*** *changes.*

Of all changes recommended in this book, this is perhaps the one of central importance. It is where the individual's needs conflict with society's authorities that the "twisted beast" of blame emerges.

Without a change in how we all respond to the inherent dilemmas in personal vs. societal health and safety standards, we merely will be cutting off one of the mythological beast's heads for each of the two that grow back to attack us. Instead, show strength and confidence in yourself and support for them in their efforts to assist you. Let them know you appreciate the changes they need to make to accommodate your special needs. As you act less and less like a victim, people will respond more positively to your requests.

It is not what I am called, but what I answer to, that shapes my image of who I am.

Unknown

This admonition for self-responsibility is not meant to be another way of saying, "Patient, heal thyself!" Rather, it is meant as a recognition that the patient, also, has a responsibility for their conduct. Your beliefs and actions will have an affect on others — your actions will influence their tendencies. And right now you are the only one available to educate those who are willing to learn.

CHAPTER SEVENTEEN

PSYCHOLOGICAL STARTING POINTS

Psychological Starting Points

The psychological and social issues tend to manifest more strongly with the more complex situations. Under the most difficult conditions, especially for those of you with a PIR of 5 or 6, they may even become so prevalent as to preclude other action until they can be stabilized.

While finding appropriate medical care, as discussed in Chapter 3, can be difficult, finding appropriate mental health care and support for chronic environmental exposure complaints is even more so. As with physicians, most have a preferred process they follow when faced with complex situations. And much depends on how familiar the mental health professional is in treating patients whose health problems defy easy explanation.

I found it nearly impossible to find mental health professionals that had even heard of these chronic conditions, let alone having a positive experience working with people suffering from them. However, I am able to present three different perspectives on the problems common to chronic ailments and some of the unique difficulties presented by environmental exposures — especially when the causes are "invisible" to others and the impact disturbs basic meanings of life such as "why me?" and "who am I now?"

The purpose of this chapter is to provide a further starting point for your own inquiry, if you find it appropriate. Like all other sections of this book, do only the minimal necessary to end your complaints. But when nothing else works, then it is often helpful to open yourself to other, previously unconsidered, ideas and systems.

Gay E. Lasher has a Doctorate in Psychology and specializes in individual and group treatment for persons with chronic medical problems and pain. She also provides treatment for mood disorders, fibromyalgia, anxiety disorders, stress management, weight management, life transitions and grief.

Carl Grimes: *How would you compare the effects of chronic environmental exposure with other chronic conditions?*

Dr. Lasher: The primary difference is the cause of the suffering. But once a person suffers, regardless of the cause, and cannot change the effects, then the psychological issues are nearly identical. These include alterations in self-perception, changes in feelings of self-efficacy, adjustment to changes in physical or mental capabilities and grieving the loss of the former self.

CG: *Does an individual's psychological issues cause illness?*

Lasher: Not in the same way that a flu virus does. Over the past few years, however, there is increasing research evidence that there appears to be communication between the central nervous system (brain and spinal cord) and the immune system. For example, psychological issues often result in stress that produces a hormone that is also present in people with depression.

CG: *Will resolving those emotional or psychological issues cure the illness?*

Lasher: No, if you're talking about illnesses such as MS or fibromyalgia. But *how* we respond emotionally and psychologically to those events can either help or hurt us in terms of lessening suffering. It's what we tell ourselves about what happened that is the key.

CG: *Can you give us an example of how this "self talk" can change our perceptions and subsequent conclusions?*

Lasher: Suppose that you are standing on a very crowded street corner waiting for the light to change. Traffic is fast and furious. People start edging closer to you, anticipating the turning of the traffic light. Suddenly, you feel a strong push from behind. You stumble a step or so out into the street before you can regain your balance, just as a truck whizzes by. What sort of feelings do you have?

CG: *First is fear. And then rage at whoever was so indifferent to my safety that they would actually push me into traffic. I'd want to push them into the traffic and I'd not care at all, at least momentarily, what happened to them.*

Lasher: Now suppose you turn around to confront your assailant and you see an elderly, frail looking person with a white cane. Instantly you realize that the person who pushed you is blind and may not have been aware of your presence until he bumped into you. How might your reaction change?

***CG**: My first reaction would still be one of fear. But I wouldn't want to vengefully harm him in return. I may even feel compassion for him and offer to help him safely cross the street.*

Lasher: Our belief about events, what we tell ourselves, can influence our eventual feelings and behavior. The activating event — the push — did occur. No matter what your emotional state or belief system, the push did occur. And the normal human immediate response is fear when there is a threat to life. However, the rest of the emotional response and potential behavior is determined by your belief or expectation at that time. If you believe the event was caused with intent to harm, you feel and act accordingly. But if you believe the act was accidental with no intent to harm, you react much differently. You may even feel compassion and a desire to help.

The sequence, as described by cognitive-behavioral theorists, is:

- An activating event
- Your belief at the time of the event
- The emotional response
- The behavioral response

You can't change the activating event, so the first step that you *can* change is your belief about the event. Emotional and behavioral responses will follow your belief.

***CG**: What are the statements that people with chronic conditions typically make?*

Lasher: The typical ones are "I can't stand it." "I'll never get better." "I feel awful." "What good am I if I can't do what I've always done?" "I don't deserve this." "Why am I being punished?" "Why won't anybody help me?"

***CG**: That sounds familiar. Don't we all talk like this at times?*

Lasher: The problem with this type of self-talk is that, even if someone is trying to help you, you won't be able to accept it as help. Your beliefs about the effectiveness of the offered suggestions and help are likely to result in a negative attitude that will push people away, further confirming your belief that no one will help you. The negative self-talk does not allow any alternative explanations for what has happened to you. It only reinforces your current belief and becomes a vicious circle.

***CG**: But isn't the alternative equally dysfunctional? If the "think correctly and heal yourself" practitioners are right, then why haven't they healed themselves? And us, too?*

Lasher: Whether or not "correct thinking," as you call it, can heal or even prevent illness is a question that is still being researched. However, I'm not talking about the physical and medical factors that cause illness. I'm talking about how we respond to what has already happened and how our thinking about that influences subsequent emotions and behaviors. Do we sit back and just wait, becoming a victim, or can we somehow find meaning in the changed situation and become a survivor?

***CG**: You're saying that we can choose? That it is our choice whether we become a victim or a survivor?*

Lasher: Yes. No matter what happens you can still choose how you are going to react. And that choice influences how you perceive subsequent events.

***CG**: What if I find it hard to give up my negativity because I firmly believe it is based on fact, on personal experience? Perhaps it has been my experience that I have been harmed, I didn't deserve it and nobody cares. Furthermore, I may not have any positive experiences or models of alternative ways of acting. How do I change my beliefs and to what do I change them into?*

Lasher: This is the hard part. To have your belief system *forcibly* changed by an accident or chronic medical condition may challenge, disrupt or even damage your current belief system. Until you can integrate the new experience and change your belief system to accommodate it, you may experience what the existentialists call "the void." You may begin to doubt who you are and what your purpose may be. You may feel victimized and very alone. Old support systems may no longer support and may even feel threatening. Basic security and survival issues arise.

To further complicate matters, a person may be so distraught and anxious that trust between patient and medical practitioner can be disrupted.

***CG**: This all sounds very bleak. Is there no hope?*

Lasher: There are many sources of hope. Edmund Whitmont*, a Jungian analyst, once wrote, "The one thing we can under no circumstances tolerate is a lack of meaning. Everything, even death and destruction, can be faced so long as it has a meaning. Even in the midst of plenty and fullness the lack of an inner sense of meaning is unbearable." The task then is to create a new meaning for what has happened and to use positively the capabilities one still has.

Literature about survivors of natural catastrophes, plane crashes and death camps as well as those who have found a way to live with chronic medical problems describe how these people had to find meaning in what had happened and then create new lives. (see Appendix B). They often transformed adversity into opportunity.

***CG**: What can one do to enhance their life in the midst of chronic suffering?*

Lasher: The specific answers to that question are different for each individual. Sometimes people who can work *with* the crisis, rather than against it, may experience a burst of creativity in some area they have ignored, become involved with a "cause," or focus on their spirituality.

David Pasikov is a therapist in private practice in Boulder, Colorado. He specializes in a new process called Body Alignment Technique, which he uses as a form of body-centered psychotherapy.

***Carl Grimes:** What happens to someone when they have an ailment that is not obvious to others, such as one they believe to be caused by an exposure to an environmental contaminant?*

David Pasikov: The ailment is not obvious because its symptoms and other attributes are not familiar to others. A common example is a person with a broken leg who requires a cast and crutches. The cast and crutches are not only readily visible but also provide an obvious and generally

---

*Whitmont, Edmund — **The Symbolic Quest,** Princeton University Press, 1978

acceptable excuse for that person's behavior deviating from the accepted standards of their family, friends and peers. Also, because the healing requirements of a broken leg are fairly well known, that person's behavior — although now much different than their peers — is fairly predictable and acceptable. No real surprises.

However, if the ailment is not visible or not immediately accepted as a legitimate excuse for not meeting common standards of performance, then that person is expected — even demanded — to stop misbehaving. If they don't, then they are assumed to be malingering — meaning that their own behavior is controllable by themselves, but they aren't willing to do so.

CG: *What affect does this have on the person with the ailment?*

Pasikov: Their self-esteem suffers and their stress level increases. The experience usually retards their recovery process because they are now focused on, among other things, meeting the expectations of others at the expense of doing what is necessary for themselves.

CG: *What happens socially?*

Pasikov: If their illness continues they will gradually lose their friends. They won't be much fun anymore. Furthermore, if they persist in their claims of illness and special need, those same "friends" often become hostile. They become intensely uncomfortable because they don't want your experience of vulnerability and mortality to reveal their own. As this process unfolds, they literally disappear and you become further alienated.

Furthermore, as you retreat to "lick your wounds," so to speak, you are also removing yourself from society. Your world becomes smaller and you increase your chances of becoming depressed.

CG: *This sounds like a stuck pattern that many of my clients describe. They also express great frustration about the lack of effective help in resolving it.*

Pasikov: It is a "revolving door" type of experience and it is usually very frustrating. I goes like this:

- We all have *life experiences* and they teach us certain things about living, about relationships, about the meaning and purpose of life itself.
- We draw *conclusions* based on those experiences.
- We develop a *belief system* whose roots are the conclusions about those life experiences.

- We then see life through the *filters* of our beliefs.
- As a result, we have *expectations* that life will work out in accordance to our beliefs.
- Finally, we complete the circle by drawing life experiences to ourselves that satisfies our expectations.

This becomes a closed-ended pursuit, allowing only those life experiences and facts that you already believe — and typically ignoring the exceptions.

CG: *Actually, that sounds very familiar, like the circularity of our understanding of public health and safety. How does one break out of this repeating pattern?*

Pasikov: I use a process that combines psychotherapy and Body Alignment Technique to access these pivotal life experiences to rethink the conclusions. Following this, new beliefs can be established. This leads to new expectations which further leads to new life experiences. As a person begins to free themselves from the revolving door of repeating patterns, they can begin to rebuild their life.

CG: *This doesn't sound like a quick-fix remedy. It sounds very complex and difficult.*

Pasikov: What is often forgotten about ourselves, and forgotten by others about their friends and family in such a life-altering situation, is that we are complex creatures living in a dynamic, ever changing world. We have a physical body, an emotional realm, a mental capacity and we are a human-being. In other words, there is a "being" or spiritual component to us as well.

But unwinding this complexity is possible. It requires motivation, curiosity, compassion, commitment and gentleness. It cannot be forced.

The most profound yet gentle tool that I have found to address the many needs at these various levels is Body Alignment Technique. This was developed by a close friend and colleague, Jeff Levin who lives in Canada (www.bodyalign.com).

As a psychotherapist, I use this process as a form of body-centered psychotherapy to emotionally and energetically support a person while in the midst of changing their fundamental belief system. Through this work,

I can assist my clients in their discovery of root causes of issues in their lives. Unresolved trauma is often held in the tissues of the body. Body Alignment Technique helps me energetically support the person in unwinding these stuck patterns.

CG: *How does that work?*

Pasikov: I approach the body as if it were a computer whose long term memory has retained past experiences of trauma, disease, pollutants, and lifestyle stress. Body Alignment Technique is a means of accessing that "computer memory" to reveal where the imbalance or blockage is located. Once located, the underlying event(s) becomes more readily available. Then it can be energetically released through balancing vibration energy-points associated with the organs, glands, and other systems of the body. When blockages are cleared from the energetic field, profound changes can then take place in the physical body.

CG: *That sounds complicated and rather esoteric.*

Pasikov: Actually, it is a practical and simple technique that is easy to learn. Because it is a vibrational system of healing — not one that requires physical or mechanical manipulation — it can be easily applied to oneself as well as to others.

CG: *What has been the experience of your patients who are highly susceptible to environmental exposures?*

Pasikov: I personally do not have a lot of experience working with severe cases of environmental sensitivities. However, I have had experience supporting people in working with allergies with positive results.

Perhaps the key to the effectiveness of Body Alignment Technique is that it has a vocabulary and an ability to work at the physical, emotional, mental and energetic levels. In other words, it addresses the multi-dimensional relationships of the presenting issue rather than assuming it is just a collection of pseudo-related physical symptoms.

Deane Shank, Ph.D. is a teacher of the Ridhwan Foundation, a school for spiritual growth. He works with his students on a one-to-one basis to develop the potential within their souls. His comments about Object Relation Theory are particularly helpful for separating the physical *causes* from both the physical and psychological symptoms; and for addressing the most difficult task of generating new meanings in your life.

Carl Grimes: *Why did you choose a spiritual-growth approach rather than just a psychological one?*

Deane Shank: The primary reason is that psychology tends to stop once the underlying psychological issue has been identified and resolved. But in my own experience, I saw that there was still more to do. For example, once I realized what was happening on a psychological level and had a cathartic release, I still had my experience of that event. What did it mean to me? How did it change other meanings? How did I feel about it? Where did I feel it in my body? What do I do now? Who am I now and who will I become? What happened to the old me?

CG: *What did an inquiry into your continuing experiences lead to?*

Shank: I was able to go beyond the boundaries of the psychological level. I was able to experience and understand both the emotional and the spiritual content of the event, which led to a deeper experience of my true nature. But what was also of interest was that the process itself; focusing on the emotions, combined with a sense of curiosity — and with no set agenda — seemed to *allow* the meaning and the spirit to be revealed. Trying to *fix* my pain or *force* the issue usually didn't work very well. But gently entering into the emotions, accompanied by large doses of compassion for myself, was more likely to lead to healing and subsequent changes.

CG: *So by using psychotherapy as a tool to achieve a higher purpose, rather than as an end in itself, you were able to move through the bleakness, depression and meaninglessness that often comes when your identity and life-meaning has just taken a major hit.*

Shank: That's right, in a simplified manner of speaking. I want to be sure we don't over simplify this process or we run the risk of missing some important subtleties.

CG: *Tell me more about the process you use.*

Shank: The process I find most helpful uses Object Relation Theory. This theory examines the relationships between a person and something else — usually another person but it can just as easily be an event such as an environmental exposure.

CG: *Could you say more about the "object" part of Object Relation Theory and what is has to do with people?*

Shank: This gets to the crux of the matter. Although our relationships are with other people, we rarely relate to a person as who he or she really is. Or for that matter, as who *we* really are.

We spend most of our life relating to that person as an object — someone or something we already know. For example, we usually relate to most women as if they were fundamentally the same as our mother — as someone we assume has the same basic characteristics. And those characteristics are *obviously* true — to us, anyway — because that is how we initially formed our knowledge of what women are like. And our experience with most other women just seems to confirm our ideas. Even when we encounter someone who doesn't conform to our expectations, we can usually just dismiss them as irrelevant or unusual — or as an aberration that ought to be avoided.

When we think of something about ourselves, we ordinarily tend to think of that particular quality as "just being me," or "this is who I am." We think about our characteristics and qualities in isolation with the rest of the world. So if I think about myself as being weak or strong or beautiful, I don't think about it as being in relation to anybody else. I see it as "just me."

CG: *That sounds like a reasonable way to define our self as our self, rather than by how others see us.*

Shank: True, if that's all there was to it. But we outgrow the original basis for our beliefs. It is important to remember that we start forming this picture of who we are in infancy. So our picture of who we are is originally based on the infantile meanings and memories of our earliest experiences — and those experiences, especially the pre-verbal ones, are primarily memories of emotions.

The experience of "remembering" the emotion contacts our current adult intelligence, discernment and wisdom. Obviously, most people would prefer to avoid re-experiencing how it felt to be a little kid in pain, and this provides perhaps the major block to resolving these issues.

CG: *Is that avoidance of reliving old trauma the reason why we tend to "objectify" people rather than "subjectify" them?*

Shank: Well, that's an interesting use for those two words, but, yes, that is one of the reasons we disconnect from people. Actually, before we disconnect from others we first distance ourselves from our own true-self. And we cannot connect to another true-self if we aren't in touch with our own.

CG: *How does the object part of our true-self form?*

Shank: As we grow and learn as infants and children, we form more ideas about our self and we begin to assemble them into a kind of picture or "image" of who we are. This collection of qualities, thoughts, feelings and experiences becomes the image of our self — our self-image. Eventually this self-image enlarges to become our identify, representing to us who we are and what we perceive our purpose in life to be. So any interference with our self-image can easily trigger an identity crisis — and the "hair trigger" for an identity crisis is any disturbance of key emotions.

But one of the fascinating characteristics of this process of assembling our self-image is that our ideas about our personality — as opposed to our true self — are actually a collection of *relationships* with other people and events who have participated in our life experiences. It forms a collection of images rather than a single picture. This collage will work just fine as long as we can keep all the separate images functioning reasonably well together.

CG: *Is that how others, and our self, become objects rather than a unique individual?*

Shank: Yes. And that describes the two components of the relationship: the self-image and our image of the object of the relationship — the object-image.

For example, if I am six feet tall and weigh 190 pounds but I see myself as small or weak, then I probably remember a time — usually subconsciously — when I saw myself in a relationship, most likely with my mother or father, who was bigger than me. And there was some kind of connection between us. It may be that I was afraid of them because of their size and they would physically hurt me. Or maybe I feel weak and helpless because they would overwhelm me with their power and authority; and I never could defeat them.

CG: *So which version is true?*

Shank: The truth is that I am big, strong and influential. And it is *also* true that I can *believe* that I am small, weak and powerless. Actually, it is also true that I can *continue* to believe that I am small, weak and powerless. That belief may follow me in my decision making processes and influence my behavior.

It is important to understand that what was true for you in infancy or childhood probably is no longer true now that you are an adult. The skills and techniques you then found necessary for survival typically continue to function into adulthood as if they are still required. Without some sort of examination of those beliefs, you will continue to act as if the world is still built around the self-image of your childhood.

CG: *How are the self-image and the object-image connected?*

Shank: There has to be a significant link between the two in order for an influential relationship to occur. The mere fact that I feel big and you feel small doesn't affect either of us. And it takes more than a chance encounter on the street to establish that connection. The connection that influences the qualities of the relationship is an emotional one — and the stronger the emotion the greater the impact. Also, the greater the emotional impact, the stronger the link.

CG: *Is the emotional link typically a single, easily identified emotion, or is it a cluster of overlapping ones?*

Shank: It is usually very complex. That's why these types of personal issues are not easily resolved. There are several steps involved in untangling this.

- Describe who you think you are. In other words, identify your self-image.
- Be aware of whatever emotions arise. Identify and experience them fully.
- The emotions will usually intensify, triggering a series of others. However, as you gain more experience you begin to realize that despite their power, the events that created them are no longer true. As you learn to see them as *past* truths, they gradually begin to dissolve and become less threatening.

- Your self-image, based upon these past events that are no longer true, begins to change and dissolve as the formative emotions lose their power.

- You then experience the dissolving of the self-image as if *you* are dissolving. Your identity is threatened — and you don't have a replacement for it yet. And as you begin to change your behavior, the "objects" of your relationships feel their emotional links being disturbed and they begin reacting to you, triggering even more of your emotional links.

CG: *This point in the process — where your very identity and purpose in life seems to be rapidly abandoning you, compounded by the hostile, or at least non-supportive, behavior of your support system — seems to be the barrier that stymies most of my highly impacted clients. It is also at this point that current medical care and mental health practices fail them. They are typically left with nothing to rely on but to return to the old patterns of the original self-image. How do they cope with that?*

Shank: By connecting to "being" rather than to "image." By experiencing what is true about yourself, despite what you were taught was true by others and from previous experience. By allowing the truth to unfold in whatever way it desires, even if you initially don't like what the truth turns out to be.

CG: *What are some of the things that clients and authorities do that either increase the harm or at least block the healing?*

Shank: It takes great care, compassion, motivation, curiosity and gentleness, on both the part of the client and the teacher, to allow healing to happen. If it is forced or hurried then that just evokes the linking emotions again. It does not lead to healing. It does not lead to any new information, new studies, new knowledge or effective guidance.

Admonitions such as the following are particularly harmful.

- Just stop it.

- Ignore it and it will go away.

- It's all in your head.

- Your exposure couldn't have been harmful.

- Don't blame me, you sabotage yourself.
- You can't be sick again so soon.
- What is it this time?.
- There is no scientific evidence that...
- There is no medical evidence...
- I followed proper procedure, so any problems are your own creation.

You need large amounts of self-love, self-compassion and self-will. But the result will be the truth.

CG: *Bring this back again to people who believe they are being exposed to something harmful in the environment. How does this type of "toxic" experience compare to being exposed to a "toxic" person?*

Shank: Disease and illness is often seen as another person, or more accurately, as another object in our collage of object relationships. As with people, we see the disease as "doing it to us," much like we see people "doing it to us."

The added burden of environmental exposure, especially when our experience is not recognized as being true, is that there is no one to respond to us. You can't address the illness as you would a person by asking questions of it and observing the response. It is silent. You don't have comparisons with how it is similar to other illnesses. It may have it's own unique patterns which are as yet undifferentiated and hidden from us.

CG: *Do you modify your procedure when working with relationships with illness rather than relationships with other people?*

Shank: Because people and illnesses are both usually related to as an object, the procedure itself doesn't need to change. However, the process is usually much more difficult. You still need to identify the patterns of the illness and how you are relating to them — especially if you have taken on the identity of the illness as part of your own identity.

It helps to label the emotions involved. The process of applying a label creates contact with the emotion. This initial general idea and experience of the pattern can often be a beginning to establishing what the emotion and the pattern means to you, how you feel about it, where you feel it in your body or your being. As you enlarge your experience you will refine the pattern and begin to identify specifics in your body. This can then lead to the origin of the pattern, parameters and individual components of your self image.

CG: *How will the actual illness be revealed ?*

Shank: As the patterns slowly reveal themselves and as the role of the illness in your self-image is discovered, then the truth of your relationship to your illness will be revealed.

Then you may be able to differentiate between emotions, mood and behavior that are triggered by a disturbance of the link between self-image and object-image, and emotions, mood and behavior directly caused by an environmental exposure.

But even if that doesn't occur, you still have a process and a support system for expanding and refining who you are and for recreating the meaning of your life.

CHAPTER EIGHTEEN

TRANSCENDING THE LOOPS

Transcending The Loops

I would like nothing better than to conclude this book with "*the* magic cure" to chronic complaints caused by non-obvious sources. Alas, that's not going to happen — not even with a standard illness that is precisely understood and expertly treated. They are still situations of influences-and-tendencies rather than cause-and-effect. They are just sufficiently "weighted" toward a more predictable outcome that they appear to be causative.

Even so, the personal impact of any condition or complaint includes much more than medicine and regulatory law. For example, a broken arm may not be catastrophic to you or me. But it would be to a professional football quarterback on the eve of the championship game — and to all those gamblers who placed large wagers based on his good health!

However, this lack of a conclusive cure is also not cause for despair. Rather it is a call for hope and for positive action. Our sufficiently advanced technology is more than capable of "seeing" what is happening with long term exposure at low levels to contaminants. We just need to ask better questions — and then fund appropriate studies for truly objective researchers — allowing our "magic" to provide better answers.

Our state of knowledge and our ability to care for others is sufficiently advanced to discern the difference between an individual's *illusion* and an *experience* that is richly unique. We just need to be willing to face our own fears and weaknesses instead of sacrificing more "necessary victims" to continue the illusion of our own special safety.

We also have the capability, both technologically and personally, of distilling the differences between true hysteria and an individual who is both chronically suffering and persistently struggling with achieving relief in the midst of a hostile "support" system. Some of the specific changes in understanding and "mind set" are:

1. The actual physical exposures and physical reactions are occurring because of "normal" substances in our habitat that we are just becoming aware of — and then stubbornly refusing to acknowledge.

2. There really isn't anything weird, unnatural — or even supernatural — about the hypersensitive individual. They are merely reacting normally to a potentially harmful situation at a level of exposure much lower than most of the population that comprises the statistical bell curve. Change the habitat — the environment, the occupants and all the interactions — and the behavior of the occupants will change. And so will the behavior of the authorities.

3. It would be useful and prudent to heed the "early warning" system that Mother Nature has so conveniently provided with the hypersensitive individual. Society would do well to use the experiences of the "front-line" scouts as a guidance system so that we all can live in peace rather than with *dis*-ease.

4. Full disclosure of products and services is imperative. If full disclosure — with the needs of individuals in mind — is not done, individuals cannot make informed choices to effectively and responsibly take care of themselves. If the primary obstacle is legal liability, perhaps the laws could be rewritten, as they have in other areas, so that true and full disclosure mitigates future liability, but failure to do so increases it. If not, the "necessary victims" of this book will also become "political victims."

5. Our scientists and medical researchers need to stop dismissing individual experience as an anomaly and become truly scientific by considering *all* the data about life on this planet. It would be helpful if they returned to the intent of the scientific method, rather than the distortions that are increasingly influenced by grant money, political pressure, managed care, legal liability and corporate "bottom lines."

6. The various disciplines that study only their own niche need to begin communicating with each other so they can integrate their knowledge and expertise. Instead of the various research factions each denying responsibility, they should work together to discover how their respective disciplines tend to influence each other. Then perhaps multi-dimensional illnesses could be successfully studied.

7. Society, public agencies and the news media could help facilitate these changes if they would stop reinforcing the "hysteria" that results when human compassion of the victim is refused — and then the victims are "driven crazy" by being accused of being "hysterical." Refocusing on the simple facts, as previously outlined, would provide more benefit with less harm to both individuals and society in general.

8. You, the ones who have been victimized, also have a job to do. Your job is the toughest because you are expected to function wisely, clearly, specifically and definitively at the very moment you physically cannot. You are the ones, because yours is no longer a shared experience, that must take control and act alone. It is you who must educate the educators. It is you who must redefine your life because others won't examine theirs. Your task is to stop behaving like a victim and set aside any hysteria, even if at that very moment hysteria may be the most "normal" behavior. You are the one that must take up the mantel of the lonely hero.

The *fundamental dilemma* between the believers and the skeptics is one of control. The instinctive human response to any non-obvious event that carries the hint of danger is either hypervigilance or denial. *Hypervigilance is a choice* of believing that "all-powerful life" imposes its demands onto the personal weaknesses of the victim. Conversely, *denial is a choice* of believing that it is a personal obligation to impose their demands—or even Divine commands — onto the weaknesses of life. Both sides have some truth. But neither belief is sufficiently true.

Each side of the dilemma is a closed loop and they don't intersect. In fact, what one group perceives as the obvious problem is precisely what the other group most easily dismisses. And what the other group perceives as the obvious problem is dismissed by the first group. The way out of any dilemma is to break out of your own loop so as to see what issues are common to both loops but whose priorities are opposite. Then each group must truthfully address that which it most fears.

The *common issue* for both groups — in the context of the dilemmas of this book — is who is in control. The *primary fear* is that someone who cannot be trusted with our best interest is in control and intending to hurt us. The *primary difference* is with how we choose to respond. The hypervigilant victims have chosen one response and the authorities of denial have chosen the opposite.

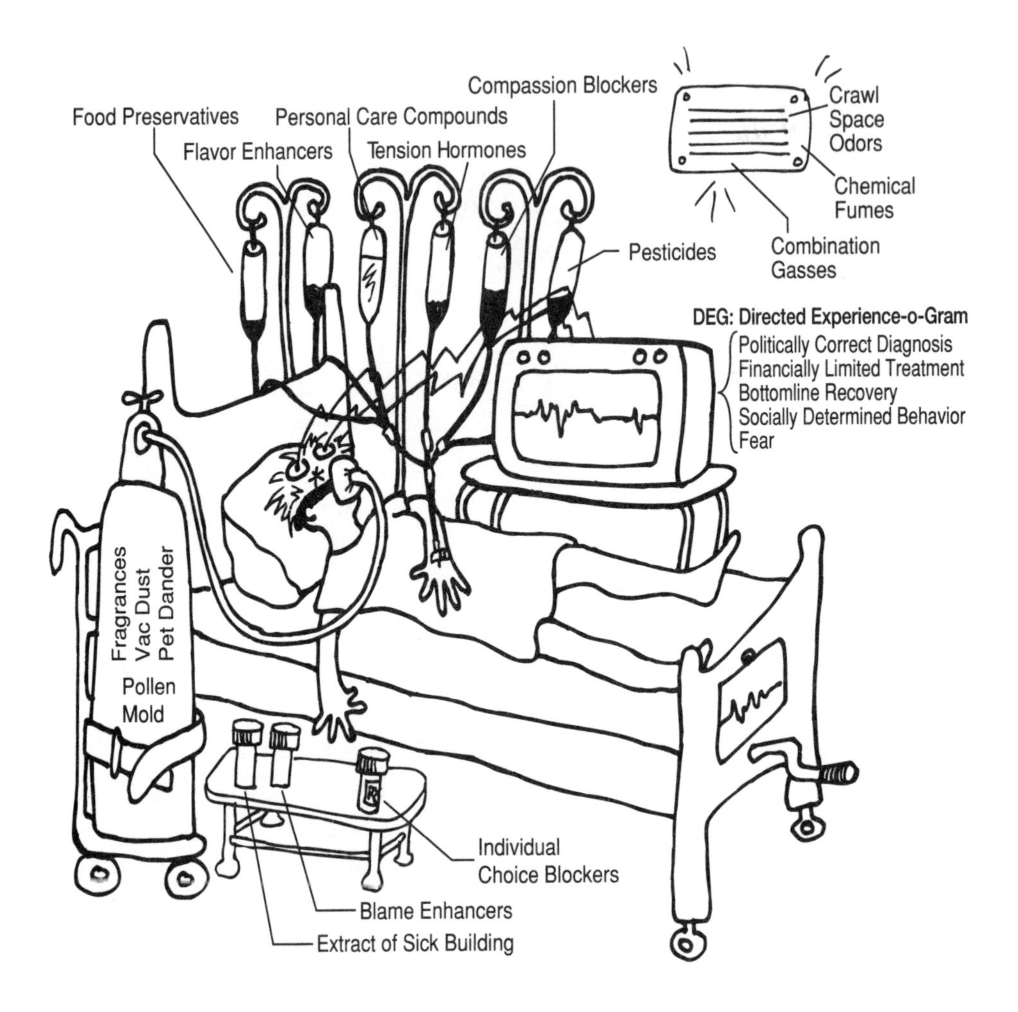

My hope is that those who have been victimized can begin to experience that which they *can* control. And that public authorities will examine their fears of that which they *can't* control. My wish is that both sides can transform their lives for the good of all, creating a new way of living.

We desperately need to forge a new alliance between individuals and authorities so that society can smoothly function, but without the abuse and victimization of the individual. This starting point requires the assumption that our individual experiences are real and that our previous efforts demonstrate our motivation and personal commitment for finding a *truthful* resolution of these non-obvious complaints.

Appendix A

Additional services and publications are available from the author's consulting company.

A booklet has been written that applies the principles of *Starting Points for a Healthy Habitat* to specific situations such as buying a new house, finding a rental, working with contractors, evaluating test results, special instructions for EPD, etc.

A video presentation is under development. The title is **Starting Points Video**. It outlines the components of the personal plan, shows how key factors and systems of houses impact your "breathing zone," and dramatizes specific examples of what actions work, which ones don't, and why.

The **Allergen Sampling Test Kit** detailed in Chapter 10.

Private Consultation with the author, Carl Grimes.

For information and prices:
Call (877) 782-7878. This is a toll-free number.
Write to GMC Media, 1811 S. Quebec Way, #99, Denver, CO 80231.
E-mail to grimes@habitats.com

See Order Form on page 287

Appendix B

Following is a diverse listing of resources. It provides a starting point with a variety of resource types. It is intended to facilitate your own information-gathering journey and to break many of the "common sense" boundaries and assumptions that limit understanding and effective action. It is not an academic or comprehensive bibliography. Many of these resources contain their own extensive bibliographies, resource guides and Internet links.

Included also are several items that not only do not recognize the types of complaints under discussion but even argue against their mere possibility. These arguments come typically from those who exclude statistical anomalies from reality. Despite that, they do have something important to say. Because the science of these illnesses has not been fully explored, not everything the individualists claim may be precisely true. Some of the so-called antagonists have found critical distinctions that need to be known and understood. Even though many of them sound as if they have an "axe to grind," it is still important to know that and to understand their arguments.

Some of the book listings have comments by me, some don't. The comments are not meant to convey added importance. Rather, they include my review of the materials and editorial comments that did not fit elsewhere in the book.

Finally, I offer numerous Internet resources. Because these can change on a daily basis, use this information as a guide. If a specific web site or discussion group is no longer accessible, use the key words as a search filter to find similar sites. To further assist you, it begins with a contents listing of the major categories.

BOOKS

Additives Book — Beatrice Trum Hunter, Keats Publishing, Inc., 1980.

This book won't include recent products. However, the fundamental information is still good. And you can look for more recent information from this author and other sources. Ms. Hunter has more than 20 books published, primarily about food and food issues.

"Advancing the Understanding of Multiple Chemical Sensitivity" — special issue of *Toxicology and Industrial Health,* Vol. 8, No. 4, 1992.

The Alchemy of Illness — Kat Duff, Pantheon Books, 1993.

A woman explores the transforming and, paradoxically, healing experience of being ill.

Allergic to the Twentieth Century: *The Explosion in Environmental Allergies — from Sick Building to Multiple Chemical Sensitivity* — Peter Redetsky, Bill Phillips (Editor), Little Brown & Company, 1997.

Allergy & Candida Cooking Made Easy — Sondra K. Lewis, with Lonnett Dietrich Blakley, 1996.

Don't let the title fool you into thinking this is just a cookbook. It is also an excellent resource guide to alternative foods, natural cooking techniques and rotational diets.

An Alternative Approach to Allergies — Theron G. Randolph, M.D., and Ralph W. Moss, Ph.D., Harper & Row, 1980.

This is the pioneering work on environmental causes of mental and physical ills. When I first started my quest for information, this was the only book that was helpful. And it is still relevant.

America Exhausted: *Breakthrough Treatments of Fatigue and Fibromyalgia* — Dr. Edward J. Conley, Vitality Press, 1998.

Dr. Conley is the founder and medical director of the Fatigue and Fibromyalgia Clinic of Michigan.

Beyond Therapy, Beyond Science: *A New Model for Healing the Whole Person* — Anne Wilson Schaef, Harper Collins, 1992.

This is a look at the limitations of science, including what it *prevents* us from accomplishing. A new model and process is offered.

Bodies in Protest: *Environmental Illness and the Struggle over Medical Knowledge* — J. Stephen Kroll-Smith, New York University Press, 1997.

This book explores how people self-describe their medical condition when none is provided by the experts. They find new meaning through their illness despite the lack of a support system.

The Body Electric: *Electromagnetism and the Foundation of Life* — Robert O. Becker, M.D., and Gary Selden, William Morrow, 1985.

Dr. Becker begins with the electrical processes most of us are familiar with — using electrical medical devices to promote the healing of broken bones — and advances to the subsequent implications. Controversial at times but well worth the read.

Brain Allergies: *The Psychonutrient Connection* — William H. Philpott, M.D., and Dwight K. Kalita, Ph.D., Keats Publishing, 1980.

The Brilliant Function of Pain — Milton Ward, Optimus Books, 1977.

The title tells the story. Instead of viewing pain as an undesirable experience that must be immediately stopped, its mere presence is what is important. It means that *something* is *causing* the pain. Find that "something" and stop the pain. Pain then becomes your guidance system for traversing the unknown terrain of your personal map rather than being held hostage by the fear of pain.

Chemical Deception: *The Toxic Threat to Health and the Environment* — Marc Lappe, Sierra Club, 1991.

Chemical Exposure and Disease: *Diagnostic and Investigative Techniques* — Janette D. Sherman, M.D., Van Nostrand Reinhold, 1988.

Although this is a very technical book, much can be learned by reading about what is tested and what is not; what is connected to disease and why.

Chemical Exposure and Human Health: *A Reference to 314 Chemicals with a Guide to Symptoms and a Directory of Organizations* — Cynthia Wilson, McFarland & Company, 1993.

Critical information for anyone with a high PIR for chemicals or who wants to prevent potential problems with children.

Chemical Exposures: *Low Levels and High Stakes* — Nicholas A. Ashford and Claudia S. Miller, Van Nostrand Reinhold, 1991.

This is the landmark study commissioned by the New Jersey State Department of Health. For the first time, authorities asked the question, "What happens when people are exposed to *below* toxic levels of chemicals." Doctors Ashford and Miller explored a very rich area of study. Their key concepts of exposures below toxic levels and their long-term effects were instrumental in understanding my clients' experiences. Their description also was more easily understood and accepted by my clients because it was congruent with their own experience. (Note also Dr. Nonas' comments in Chapter 3 about congruence between patient beliefs and medical beliefs.)

Chemical Sensitivity: *The Truth About Environmental Illness (Consumer Health Library* — Stephen J. Barrett and Ronald E. Gots, Prometheus Books, 1998.

Read this book to learn how the chemical industry defends itself against claims of harm from their products.

Chronic Fatigue & Tiredness: *Effective Solutions for Conditions Associated with Chronic Fatigue Syndrome, Candida, Allergies, PMS, Menopause, Anemia, Low Thyroid, and Depression* — Susan M. Lark, M.D., Westchester Publications, 1993.

A Civil Action — Jonathan Harr, Vintage Books/Random House, 1995.

This book is a spellbinding, nonfiction account of one of the first major court cases claiming that a chemical manufacturer was responsible for contaminated drinking water that caused a variety of illnesses. Not only is the legal and political intrigue captivating, there is a wealth of information about why legal remedies are so difficult to obtain, who has data from epidemiological studies, and the correlation of neurotoxicity testing with chemical exposures. One specific finding was that drinking contaminated water was not the primary source of exposure. Bathing and showering were additional exposures and they most likely posed even greater health risks than the drinking water. A major motion picture by the same title and based on this book has recently been released.

Clean and Green: *The Complete Guide to Nontoxic and Environmentally Safe Housekeeping* — Annie Berthod-Bond, Ceres Press, 1990.

If you can't find cleaning products that are safe for you, get this book and learn how to make your own. Don't be deceived by the book's small size. It's a simple concept with lots of specific formulas that work extremely well.

Clearer, Cleaner, Safer, Greener: *A Blueprint for Detoxifying Your Environment* — Gary Null, Villard Books, 1990.

Good fundamentals for how to remove most common sources of exposure.

The Coming Plague: *Newly Emerging Diseases in a World Out of Balance* — Laurie Garrett, Farrar, Straus and Giroux, 1994.

Other similar books received more publicity and one was made into a movie. Written by an award-winning journalist, this book is a thorough documentation of the diseases and why they are emerging. Understanding infectious disease epidemics on a world basis underscores why public health officials consider individual illness as a very low priority. That knowledge could also help individual sufferers to better communicate their needs in a different, more appropriate way.

Common-Sense Pest Control: *Least Toxic Solutions for Your Home, Garden, Pets and Community* — William Olkowski, Sheila Daar, and Helga Olkowski, The Taunton Press, 1991.

You *don't* have to spray toxic chemicals once a month or on any other contractual schedule to effectively control pests. In fact, you don't have to spray toxic chemicals at all. Read this book to find out how to avoid spraying. Also, read anything you can find on Integrated Pest Management (IPM).

Cross Currents: *The Perils of Electropollution / The Promise of Electromedicine* — Robert O. Becker, M.D., Jeremy P. Tarcher, 1990.

Dr. Becker continues his exploration from *The Body Electric,* plus advances his own theory about how electromagnetic fields affect the human body.

Culture of Denial: *Why the Environmental Movement Needs a Strategy for Reforming Universities and Public Schools* — C.A. Bowers, SUNY Press, 1997.

Currents of Death: *Power Lines, Computer Terminals and the Attempt to Cover Up Their Threat to Your Health* — Paul Brodeur, Simon and Schuster, 1989.

I have not talked about electromagnetic exposure in this book. However, if it is of concern to you, Mr. Brodeur's book is a great place to start your education.

Defining Multiple Chemical Sensitivity — Bonnye L. Matthews (Editor), McFarland & Company, 1998.

One of the best of many excellent books, with careful documentation, that answers most of the charges of the "anti" group. Documentation of evidence pointing to the reality of MCS includes the latest medical studies using the most advanced technology. Also included are sections on workers' compensation procedures, toxic torts litigation and separating fact from "spin" fiction in the literature.

Detoxification & Healing: *The Key to Optimal Health* — Sidney MacDonald Baker, M.D., foreword by Jeffrey Bland, Ph.D., 1997.

The Dispossessed: *Living with Multiple Chemical Sensitivities* — Rhonda Zwilinger, text and photographs, The Dispossessed Project, PO Box 402, Paulden, AZ 86334-0402, 1998. E-mail: rzdisp@northlink.com.

What would it be like to have a generalized susceptibility so far to the left of the bell curve that you cannot tolerate any level of exposures to even the most common substances? What happens if you can find no safe place to live? Where do you go? How do you not just cope, but *survive*?

In a sense you become homeless, because you are totally separated from your family and friends and have no safe place to live. But instead of living under a bridge or in a cardboard box in an alley — because you can't tolerate to breathe the polluted air — you find the deserts of the southwestern U.S. For some, the deserts are sufficient. But for others, it isn't. They can't be around other people, even ones like themselves, because what is safe for others is not safe for them.

The dispossessed are people so severely impacted by such extremely low levels of exposure that they can survive only under the most pristine conditions. While this impact level may never apply to you (hopefully!), it is important to know that it exists so you can accurately determine your condition between the extremes of no impact and total impact.

If your PIR is this severe, you need to know about this option — and that you are not totally alone. And if your PIR is a 5 or less, then count your blessings, express your gratitude and take action to care for yourself. Then get busy helping and educating others.

The Diving Bell and the Butterfly — Jean-Dominique Bauby, Alfred A. Knopf, 1997.

This is an autobiographical account by an energetic, bright, extremely successful man who was struck suddenly with a deadly illness. He could perceive perfectly and he could think and remember flawlessly, but was incapable of speech or most other means of communication. He wrote this book with a special computerized method of writing that selected one letter at a time. His story is all the more tragic because he died just prior to its publication.

The Dose Makes the Poison: *A Plain-Language Guide to Toxicology* — M. Alice Ottoboni, Ph.D., John Wiley & Sons, 1997.

An excellent review of chemical health-hazards for the layman. The book discusses the dose-response relationship of chemicals in the air, food and water for determining what is harmful and what is harmless. It is important to understand the dose-response relationship. It is also important to realize that it describes the bell curve measurements of large groups of people, not individual reactivity.

Drinking Water Hazards: *How to Know if There Are Toxic Chemicals in Your Water and What to Do if There Are* — John Cary Stewart, Envirographics, 1990.

Economic Architecture — Richard L. Crowther, FAIA, Butterworth Architecture, 1992.

Excellent material not only about architectural factors in healthy indoor habitats but also discussion about what makes up a healthy habitat. Crowther was a pioneer in solar heating and has actively investigated the role of positive and negative ions. He currently lives in a house of his own design that most hypersensitive people tolerate quite well.

Environmental Illness: *Myth and Reality* — Herman Staudenmayer, Lewis Publishers, Inc. 1998.

Another controversial book, at least as far as chronic sufferers are concerned. Herman Staudenmayer believes that environmental illness is based more on fear than by any actual exposures to toxins. He claims to have successfully helped hundreds of patients in nearly 20 years of treatment by dismissing much of the information that, as he describes it, defends a culture of victimization. While I personally disagree with most of his claims and how others use them in a dismissive manner, it is still important to be aware of them. I also consider it critical to discuss more fully the issues of victimization in terms of stopping exposure events, eliminating the harm from abusive medical relationships, and in terms of making choices about responding to those events.

Environmental Testing: *Where To Look, What To Look For, How To Do It, and What Does It Mean?* — Citizens Clearing House for Hazardous Waste, P.O. Box 6806, Falls Church, VA 22040.

Ethics on Call: *A Medical Ethicist Shows How to Take Charge of Life-and-Death Choices* — Nancy Dubler, Esq, and David Nimmons, Harmony Books, 1992.

I constantly emphasize the importance — even the necessity — of making personal choices to take charge of your life. Here's a "guidebook" to assist you.

Evaluate Your Own Biochemical Individuality (Self-Care Health Library Series) — Jeffrey S. Bland, Ph.D., Keats Publishing, 1987.

5 Years Without Food: *The Food Allergy Survival Guide* — Nicolette M. Dumke, Allergy Adapt, Inc., 1997.

If you have allergies or other intolerance to foods, get this book before you starve! This is more than just a recipe book. It's a wonderful resource for basic information on food intolerance, alternative foods, food substitution, rotation diets, food classification guidelines and measurement equivalencies. For example, if a standard recipe calls for 2 cups of wheat flour, what amount of quinoa or spelt flour would you use? This book should be a "bible" for those of you on EPD treatments.

Flow: *The Psychology of Optimal Experience/Steps Toward Enhancing the Quality of Life* — Mihaly Csikszentmihalyi, Harper & Row, Publishers, 1990.

Harmony and meaning in life from merely existing? Regardless of the size of your bank account, your social standing or the state of your health? Read this book along with those of Frankl and Weisel.

The Four Pillars of Healing: *How Integrated Medicine Can Heal You* — Leo Galland, Random House, 1997.

Conflict between an individual and the environment is the fundamental health problem of our times. Traditional medicine performs poorly in this arena because it rarely recognizes the individual. It tends to see the people as statistical bell curves. The closest it comes to acknowledging a person as an individual is to label them by their symptoms. "He's the heart case, she's the liver trauma, that kid's the broken leg." Is it any wonder people are looking elsewhere for help? If you are, then start here.

The Greatest Benefit to Mankind: *A Medical History of Humanity* — Roy Porter, Norton, 1997.

The Green Kitchen Handbook: *Practical Advice, References and Sources for Transforming the Center of Your Home Into a Healthy, Livable Place* — Annie Berthold-Bond, Harper Collins, 1997.

Another great how-to book and resource guide from Annie Berthold-Bond. Very comprehensive from the simple tips to the complex projects; from how to comprehend the truth of food labels to how to mix and match alternatives to fit your individual needs.

Guns, Germs and Steel — Jared Diamond, Norton, 1998.

If you have any doubts that the physical environment can affect people, or in this case the whole of history, read this Pulitzer Prize winning book.

The Healing Arts: *Exploring the Medical Ways of the World* — Ted Kaptchuk and Michael Croucher, Summit Books, 1987.

A fascinating global survey of the unique cultural practices of medicine around the world and how they relate to our beliefs in the authority of our own medical culture. Imagine trying to comply with the instructions of a "witch doctor?" — or him complying with a modern technologist. The story of this book is how culture and technology were successfully merged at one healing center.

Healing Environments: *Your Guide to Indoor Well-Being* — Carol Venolia, Celestial Arts, 1988.

Healing Nutrients: *The People's Guide to Using Common Nutrients That Will Help You Feel Better Than You Ever Thought Possible*–Patrick Quillin, Ph.D., R.D. Contemporary Books, 1987.

Healing Words: *The Power of Prayer and the Practice of Medicine* — Larry Dossey, M.D., Harper Collins, 1993.

Larry Dossey is one of the leading proponents of alternative methods of healing. And he documents his claims with the latest science, including double-blind controlled studies. Recent studies on how mental attitudes affect human organs and cells give hope to the possibility that a diagnosis of "it's all in your head" leads to a specific treatment plan, rather than being used as a dismissive insult.

Healthful Houses: *How to Design and Build Your Own* — Clint Good with Debra Lynn Dadd, Guaranty Press, 1988.

Healthy House Building — John Bower, John Stuart Lyle, 1989.

Bower's books are based on his experience of making his own house safe for his wife. A very reliable source; excellent information.

The Healthy School Handbook: *Conquering the Sick Building Syndrome and Other Environmental Hazards In and Around Your School* — Norma L. Miller, Ed.D., Editor, NEA Professional Library, PO Box 509, West Haven, CT 06516, 1 (800) 229-4200.

The list of authors is a veritable Who's Who of leaders on the indoor environmental exposure front. And the focus is on the one area of indoor exposure that affects the most vulnerable spectrum of the population — our children. In school situations it is almost impossible to obtain any action, due to perceived priorities and failed bond elections.

Help for the Hyperactive Child: *A good-sense guide for parents of children with hyperactivity, attention deficits and other behavior and learning problems* — William G. Crook, M.D., Professional Books, 1991.

The Highly Sensitive Person: *How to Thrive When the World Overwhelms You* — Elaine N. Aron, Carol Publishing Group, 1996.

How to Grow Fresh Air: *50 Houseplants that Purify Your Home or Office* — Dr. D. C. Wolverton, A Penguin Book, 1996.

NASA wanted to create a lunar habitat that provided oxygen without mechanical equipment. Dr. Wolverton was one of the researchers. This book applies their experience to the indoor habitat.

Hystories: *Hysterical Epidemics and Modern Media* — Elaine Showalter, Columbia University Press, 1998.

This book has created its own form of hysteria. Readers are typically polarized by its findings. For example, at a book signing I attended, many in the audience were warmly appreciative of Showalter's revelations about cults and how the media covers them. On the other hand, she was challenged by a representative of a local chronic fatigue support group for including CFIDS in the same hysterical category as alien abductions.

Therefore, it is an excellent source for experiencing the clash of the various "truths" as defined by science, regulatory agencies, academic research, personal experience and the media (which includes this book). Again, even if you totally disagree with her findings, it is important to understand the issues in order to defend yourself and to educate others without becoming hysterical in the process.

The Illness Narratives: *Suffering, Healing, and the Human Condition* — Arthur Kleinman, M.D., Basic Books, Inc. 1988.

The Impossible Child: *In School, At Home* — Doris Rapp, M.D., Practical Allergy Research Foundation, PO Box 60, Buffalo, NY 14423, 1986.

Indoor Air: *Risks and Remedies* — Richard Crowther, FAIA.

Out of print, but worth looking for in your local library.

Invisible Trauma: *The Psychosocial Effects of Invisible Environmental Contaminants* — Henry M. Vyner, M.D., Lexington Books, 1988.

This is the cornerstone for the basic methodology I have developed. As evidence of the correctness of Dr. Vyner's study, I offer the dramatically increased success rate of my clients when I applied his findings as an alternative to their conventional belief systems. And when I taught his model of Denial-Vigilance-Hypervigilance to my clients, they were better able to understand and manage the nonphysical aspects of their situation.

Dr. Vyner also discusses the issue of hypochondria, how it is diagnosed, and the accuracy of such diagnosis. He concludes that misdiagnosis of hypochondria usually will result in more harm to the patient than from environmental exposure itself. He then concludes with some sage advice for the public authorities who must deal with those individual events.

Is This Your Child's World? *How You Can Fix the Schools and Homes That Are Making Your Children Sick.* — Doris J. Rapp, M.D., FAEM, FAAP, FAAA, Bantam Books, 1996.

This is a critical book for any parent, teacher, or school administrator. Some of my clients'children have been evaluated as "learning disabled," only to be reevaluated as "gifted-and-talented" after exposure sources were removed or isolated.

Jet Smart–Diana Fairechild, Flyana Rhyme, 1992.

This is her personal story about how she was exposed to pesticides as an airline attendant on commercial aircraft, and what she did about it. Be sure to check out her Web site at <www.flyana.com> for comprehensive information about health and safety on commercial airlines.

Lead and Your Drinking Water — EPA booklet, (202) 260-2080.

Just a small part of extensive information from the U.S. Environmental Protection Agency. (Also see the EPA Web site.)

Living Downstream: *An Ecologist Looks at Cancer and the Environment* — Sandra Steingraber, Perseus Print, 1997.

Living with Environmental Illness — Stephen Edelson and Jan Berliner Statman, Taylor Publishing, 1998.

The Man Who Tasted Shapes: *A Bizarre Medical Mystery Offers Revolutionary Insights into Emotions, Reasoning, and Consciousness* — Richard E. Cytowic, M.D., Jeremy P. Tarcher/Putnam Books, 1993.

Although not directly related to the topic under discussion, it is an interesting study of how our nervous system and brain can sometimes *slightly* malfunction. This is all the more relevant now that the current technology of PET scanners have made visible the direct effect of certain odors on the emotional control centers of the human brain.

Man's Search for Meaning — Viktor E. Frankl, Beacon Press (paperback 4th Edition), 1992. Original U.S. copyright 1959.

This is necessary reading for anyone who has had his or her life's belief system disrupted or destroyed.

Medicine on Trial: *The Appalling Story of Ineptitude, Malfeasance, Neglect, and Arrogance* — Charles B. Inlander, Lowell S. Levin and Ed Weiner, Prentice Hall Press. Copyright 1988 by The People's Medical Society.

Evidence for the authors' claims come from the medical profession's own studies as reported in their own professional journals.

Multiple Chemical Sensitivity: *A Scientific Overview* — Frank L. Mitchell, Editor, U.S. Department of Health and Human Services, Public Health Service, Agency for Toxic Substances and Disease Registry (ATSDR), 1995.

Important collection of a variety of authoritative opinions.

Molecules of Emotion: *Why You Feel the Way You Feel* — Candace B. Pert, Ph.D., Scribner, 1997.

A fascinating story of Dr. Pert's discoveries and experiences as a leading research neuroscientist, including her part in the discoveries of the role of peptides and their receptors as a possible connection between the physical and mental components of life.

Candace Pert became a hero of mine when I read an interview about 15 years ago where she asserted that the immune system may well be a sensory organ. In fact, it fit the definition of perception at least as well as those of the standard five senses.

Her book further increases my curiosity about some of the latest approaches to the interconnection of all the major systems of the human organism, including the mind. Dr. Pert mentions several experiments and peer-reviewed reports which demonstrate that there is more to the human brain than just the organ contained within the skull. The latest findings, based on peptides and their receptors, seem to indicate that "brain-like" processing actually occurs throughout the *whole* body.

If true, it begs the question, "If the whole body is the brain, what are the effects of environmental exposures — especially chemicals that are absorbed into and even penetrate the barrier of the skin — on the physical functioning of the body, our sense of well-being, and on our mood and behavior?r?

Finally, her 20-page appendix of resources and practitioners of BodyMind Medicine, as she refers to it, is alone worth the price of the book.

Molecules of the Mind: *The Brave New Science of Molecular Psychology* — Jon Franklin, Atheneum, 1987.

This story of the pharmaceutical effects on people is riveting. It fits well with the Pert book described above. Particularly memorable was his description of how a *single* molecule from a female gypsy moth was sufficient to activate the entire complex mating behavior of a male gypsy moth. If a *single* molecule is sufficient, then how relevant is the standard description of exposure based on measurements of concentration? If a single molecule is sufficient to cause an effect, then parts-per-million, even parts-per-billion, may be grossly insufficient and misleading.

Natural Detoxification: *The Complete Guide to Clearing Your Body of Toxins* — Jacqueline Krohn, M.D.; Frances A. Taylor, M.A.; and Jinger Prosser, L.M.T.; Hartley & Marks, 1996.

The Natural House Book: *Creating a Healthy, Harmonious, and Ecologically-Sound Home Environment* — David Pearson, Fireside, Simon and Schuster, 1989.

This has been one of my favorite books since I first discovered it in 1991. Not only does Pearson do a complete evaluation of what to remove from a house to make it more healthy, he puts equal weight on what to put *into* it to make it esthetically pleasing and harmonious with your life preferences. In other words, people with a high impact rating (PIR) don't have to be confined to living in a sterile "bubble."

The Natural House Catalog: *Everything You Need to Create an Environmentally Friendly Home* — David Pearson, Fireside, Simon and Schuster, 1996.

This logical extension of his previous book (above) is a "must have" book in anyone's library. Also, I am honored to be featured on page 120.

The Night Trilogy: *Night, Dawn,* and *The Accident* — Elie Wiesel, Hill & Wang, 1972, 1985.

This paperback edition is a collection of the three original works entitled *Night, Dawn* and *The Accident,* another indispensable source of knowledge, inspiration and hope for those of you with extreme PIRs who have had your belief systems damaged or destroyed.

Nontoxic & Natural: *How to Avoid Dangerous Everyday Products and Buy or Make Safe Ones* — Debra Lynn Dadd, Jeremy P. Tarcher, 1984.

All books by Debra Lynn Dadd are "must have" books.

The Nontoxic Home: *Protecting Yourself and Your Family from Everyday Toxics and Health Hazards* — Debra Lynn Dadd, Jeremy P. Tarcher, 1986.

Nontoxic, Natural and Earthwise: *How to Protect Yourself and Your family from Harmful Products and Live in Harmony with the Earth* — Debra Lynn Dadd, Jeremy P. Tarcher, 1990.

Nutritional Influences of Illness: *A Sourcebook of Clinical Research* — Melvyn R. Werbach, M.D., Keats Publishing, Inc., 1988.

The information in this book transcends the claim of "we are what we eat." It would be interesting to reread this book in light of the information in Candace Pert's book, *Molecules of Emotion.*

Occupational Medicine: *Workers with Multiple Chemical Sensitivities* — Mark Cullen, M.D., Editor, Vol 2/Number 4 October-December 1987, Hanley & Belfus, Inc.

This offers an overview of the competing hypotheses of what causes multiple chemical sensitivity in the workplace. Dr. Cullen's conclusion is that although none of them can be proven, people who continue to suffer demand at least basic human respect and the continuing best efforts of the medical profession to discover what is happening rather than merely dismissing such patients as "troublesome hypochondriacs."

Our Children's Toxic Legacy: *How Science and Law Fail to Protect Us from Pesticides* —John Wargo, Yale University Press, 1998.

Our Stolen Future: *Are We Threatening Our Fertility, Intelligence and Survival* — Theo Colburn, Dianne Dumanoski and John Myers, Dutton, 1996.

This is another must-read book. It discusses the current evidence that synthetic chemicals affect life in more ways than by causing cancer or altering our genetics. It discusses the evidence that points to how these chemicals affect the function and communication *within* our bodies. The internal systems include our ability to reproduce at all, not just without mutation; to think, consider, evaluate and choose clearly, with wisdom; and to protect ourselves internally from outside dangers with our immune system.

Out of This World: *A Woman's Life Among the Amish* — Mary Swander, Viking, 1995.

The main story is about her experience among the Amish in Iowa. However, the reason she chose this life-style and geographic location makes this book very relevant. Mary Swander suffers from Environmental Illness. Her book is about her individual journey, how she created her own map of discovery and how she is now using that experience to help others. When I met her at a book signing, I saw her as one of those rare people who was not only surviving, but was now *thriving* in the midst of her forcibly altered life.

Perceiving Ordinary Magic: *Science and Intuitive Wisdom* — Jeremey W. Hayward, New Science Library, 1984.

Pesticides and Human Health — William H. Hallenbeck and Kathleen M. Cunningham-Burns, Springer-Verlag, 1985.

Poisoning Our Children: *Surviving in a Toxic World* — Nancy Sokol Green, The Noble Press, 1991. Foreword by Sherry Rogers, M.D.

Power Healing: *Use the New Integrated Medicine to Cure Yourself* — Leo Galland, Random House, 1998.

Prescription for Nutritional Healing: *A Practical A-Z Reference to Drug-Free Remedies Using Vitamins, Minerals, Herbs & Food Supplements*–James F. Balch, M.D., and Phyllis A. Balch, C.N.C., Avery Publishing Group, 1997.

Remarkable Recovery: *What Extraordinary Healings Tell Us About Getting Well and Staying Well* — Caryle Hirshberg and Marc Ian Barasch, Riverhead Books, 1995.

Sixty people who had "spontaneous remissions" were interviewed for the purpose of discovering what they did differently. A key issue is whether psychological and social behaviors merely comfort the patient or whether they have actual healing effects.

The Secret House: *24 Hours in the Strange and Unexpected World in Which We Spend Our Nights and Days* — David Bodanis, Touchstone, Simon & Schuster, 1986.

Not only is this a fun book about eggs that breathe, wind storms in our pant legs and lightning in our polyester shirts, it changes the scale of how we perceive what happens in our house. A great exercise if you get stuck in understanding your own home's effect on you.

Shame: *The Power of Caring* — Gershen Kaufman. Schenkman Books, Inc., 1985, 2nd Edition Revised.

Silent Spring — Rachel Carson, Houghton Mifflin Company, 1962.

Return to the original warning about how the fouling of our own nest is endangering our wildlife. However, realize that what is true for the wildlife of this planet is also true for us. We are not a separate, privileged life-form that is protected automatically from what harms the rest of the earth.

Sinus Survival: *The Holistic Medical Treatment for Allergies, Asthma, Bronchitis, Colds, and Sinusitis* — Robert S. Ivker, D.O., Jeremy P. Tarcher/Putnam, 3rd Ed. 1995.

Chronic sinusitis has been the nation's number-one chronic disease for over the last decade. That fact illuminates two key points: There has to be a fundamental difference between the medical care of acute illness and chronic illness. Acute illness is "fixable." Chronic illness, by definition, cannot be cured. That is why it continues to be present and is called "chronic." Since it cannot be cured, the best intervention is prevention. Dr. Ivker presents an integrated, whole person, approach. This should be the "bible" for sufferers of chronic sinusitis and other chronic respiratory illnesses. Dr. Ivker is a two-time president of the American Holistic Medical Association.

Spirit of Survival — Gail Sheehy, William Morrow and Company, Inc., 1986.

This is the story of the suffering, struggle and eventual triumph of a young Cambodian woman. Pol Pot destroyed her world when she was only 6 years old. But then the positive changes wrought by coming to America were also disruptive. This is more than just a story about *survival*. It explores the qualities of people who transform their victimization into *thriving*.

The Social Transformation of American Medicine: *The Rise of Medical Authority and the Shaping of the Medical System* — Paul Starr, Basic Books, Inc., 1982.

This won the 1984 Pulitzer Prize for General Nonfiction. The "landscape" of the medical system has since been further altered by managed "care."

Staying Well in a Toxic World: *Understanding Environmental Illness, Multiple Chemical Sensitivities, Chemical Injuries and Sick Building Syndrome* — Lynn Lawson, The Noble Press, 1993. Foreword by Theron G. Randolph, M.D.

Superimmunity for Kids — Leo Galland, M.D., Dian Dincin Buchman, Delacorte Press, 1989.

This is a "must have" nutrition book about childhood and pregnancy. Kids are different. They are not identical to adults except for their smaller size. Doctors, medical researchers and regulatory agencies have accepted this for years with their warning that a drug is safe, except for more vulnerable people such as the elderly and children. Develop your children's healthy life-style from the start.

Surviving Modern Medicine: *Finding the Right Cure, Making Informed Choices about Conventional and Alternative Medicine* — Frances Taylor and Jacqueline Krohn, Hartley & Marks, 1998.

Take This Book to the Hospital With You: *A Consumer Guide to Surviving Your Hospital Stay* — Charles B. Inlander and Ed Weiner, People's Medical Society, Rodale Press, 1985.

This tells how to create a personal support system for when you are at the mercy of the medical "machine."

Textbook of Clinical Occupational and Environmental Medicine–Linda Rosenstock and Mark Cullen, W B Saunders Company, 1994.

They Say You're Crazy: *How the World's Most Powerful Psychiatrists Decide Who's Normal* — Paula J. Caplan, Ph.D., Addison-Wesley Publishing Company, 1995.

This is the inside story of the DSM.

Tired or Toxic: *A Blueprint For Health* — Sherry A. Rogers, M.D., Prentice Publishing, 1990.

A must-read for anyone with symptoms of unrelieved fatigue.

The Toxic Cloud — Michael H. Brown, Harper & Row, 1987.

This is primarily about the outdoor environment, but it is full of insights about how chemical and other exposure sources are circulated by air movement around the globe, and how small quantities can accumulate thousands of miles away in the most unanticipated locations.

NOTE: Eleven years after publication of this book, there was an Associated Press article about how scientists recently confirmed that pollution from China and Central Asia has traveled as far as Texas. The same article also stated that several U.S. locations showed that pollution from these sources increased to within 2/3 of federal health standards.

Toxics A to Z: *A Guide to Everyday Pollution Hazards* — John Harte, Cheryl Holdren, Richard Schneider, and Christine Shirley, University of California Press, 1991.

The presence and locations of most of these toxins will surprise many of you. This book helps to confirm the point that common objects often contain substances that in other contexts are considered toxic, which raises the question of why they are considered safe in these.

The 20-Day Rejuvenation Diet Program: *With the Revolutionary Phytonutrient Diet* — Jeffery S. Bland, Ph.D., Sara H. Benum (Contributor), Keats Publishing, 1996.

Waking the Tiger: *Healing Trauma: The Innate Capacity to Transform Overwhelming Experiences* — Peter A. Levine and Ann Frederick, North Atlantic Books, 1997.

This book fits nicely with Dr. Pert's *Molecules of Emotion.* Animals face life-and-death events every day. Why aren't they continually *traumatized*? I was taught that animals are "dumb," meaning they aren't intelligent enough to even feel pain in the same way people do. If they can't feel pain then they can't be traumatized. Is trauma a condition of our advanced state as a life-form?

Peter Levine argues that animals do experience the pain of their struggles, but they have an innate way of healing them. If this is true, then why don't people? We are, after all, superior to animals, aren't we? Doesn't our very history demonstrate time after time, struggle after struggle, that we are above animals in every way? Even separate from them? We train ourselves to act differently than animals. We observe what animals do and that is what we are ***not*** to do.

Perhaps that is the one of the reasons that people rarely heal their emotional trauma. By separating ourselves from nature in every possible way, in order to establish our superiority, we have disconnected from our innate capabilities of healing. Actually, I believe it's more likely to be, using Pert's peptide and receptor system, that we have *disrupted* the innate functioning of our MindBody, precluding self-healing. Perhaps miraculous healings like those detailed in the book *Remarkable Recovery* are actually more natural than remarkable.

The Whole Way to Allergy Relief & Prevention: *A Doctor's Complete Guide to Treatment & Self-Care* — F. Taylor, E. Larson, J. Krohn, J. Prosser (Contributor). Hartley & Marks, 1997.

This book includes *both* traditional medicine and holistic medicine for explaining our bodies defenses, management of stress, life-style and nutrition, immunotherapy and detoxification methods.

The Whole Way to Natural Detoxification: Clearing Your Body of Toxins–Frances Taylor, Jinger Prosser, Jacqueline Krohn, Hartley & Marks, 1996.

Why Your House May Endanger Your Health — Alfred V. Zamm, M.D., with Robert Gannon, Touchstone, Simon and Schuster, 1980.

This book is one of the first warnings about indoor exposures.

Your Home, Your Health, and Well-Being — David Rousseau, W.J. Rea, M.D., and Jean Enwright, Ten Speed Press, 1988.

One of the first complete books about sources of exposure and what to do about them.

The Yeast Connection: *A Medical Breakthrough* — William G. Crook, M.D., Professional Books, 1983.

This was Dr. Crook's first book about the "missing diagnosis" — how there is more to infectious agents than just bacteria and viruses. Overgrowth of yeast organisms has its own set of symptoms, treatment and complications.

The Yeast Connection Handbook — William G. Crook, M.D., Professional Books, 1997.

This updates the original best-selling *The Yeast Connection.* It covers such topics as the relation of common yeast overgrowth to PMS, asthma, digestive and urinary problems, sexual dysfunction, psoriasis, multiple sclerosis and muscle pain.

The Yeast Connection Cookbook: *A Guide to Good Nutrition and Better Health* — William G. Crook, M.D., and Marjorie Hurt Jones, R.N., Professional Books, 1997.

An excellent companion to the other yeast connection books.

Internet

The Internet can be both wondrous and horrible — wondrous in that there is more information available at the click of a button than at any time or place in the history of mankind. The horrible aspect is that you don't always know the reliability of the information. Misinformation, outright lies and institutional or industrial "spin" are rampant — all the more reason to find as much diverse information as you can so your decisions are the best they can be.

This section is organized into several categories and some of the Web sites appear in more than one category. I assume a minimal working familiarity with the Internet, Web pages, E-mail and Usenet Discussion Groups.

Also, please realize that the dynamic nature of the World Wide Web will make some of this information obsolete by the time this book gets to the printer. For that reason, become very familiar with the various search engines that are generally available and with the more specific ones in this resource guide. Search for key words and phrases with different search engines. The variety of responses is always surprising. And sometimes the one with the "worst" results contains a "gem" hidden somewhere in the 200th response. Explore the available links at each site and the "What's Related" button on your browser (if it has one). When you find that "gem," let others know about it!

Web Site Contents

ACADEMIC SITES

Most of these sites appear under other categories.
However, many people want academic resources.

Columbia University
http://www.columbia.edu/cu/healthwise/
Includes an advice column, "Go Ask Alice." Answers are from health professionals.

Cornell University
http://pmep.cce.cornell.edu/
Pesticide Management Education Program (PMEP)
- also - Pesticide Active Ingredient Profiles.
http://pmep.cce.cornell.edu/profiles/index.html

Emory University
http://WWW.MedWeb.Emory.Edu/MedWeb/
Search engine for Alternative Medicine Sites.

Hamline University.
http://www.hamline.edu/lupus
A list of articles about lupus with links to other areas.

Newcastle University
www.newcastle.edu.au/department/bi/birjt/cpruis/urine.html
Characteristic anomalies in CFS urine profiles.

University of California, Irvine Health Promotion Center
http://www.socecol.uci.edu/~socecol/depart/research/hpc/hpc.html

University of Iowa College of Dentistry.
http://indy.radiology.uiowa.edu/Beyond/Dentistry/sites.html

University of Iowa Institute for Rural and Environmental Health
http://info.pmeh.uiowa.edu

University of Kansas School of Allied Health
http://www.kumc.edu/SAH/

University of London Ergonomics and Human Computer Interaction
http://www.ergohci.ucl.ac.uk/

University of Pittsburgh
http://www.pitt.edu/~cbw/altm.html

Alternative Medicine HomePage
Links to unconventional, unorthodox, unproven, or alternative, complementary, innovative, integrative therapies.

University of Utah
gopher://atlas.chem.utah.edu:70/11/MSDS

AIRLINE TRAVEL

Healthy Flying with Diana Fairechild
http://www.flyana.com/
Comprehensive information about healthy air and other issues when flying commercial airlines. Includes instructions on how to request "full air utilization" in the cabin. This web page is by the author of the book *Jet Smart,* Diana Fairechild's personal story of illness caused by un-notified pesticide use on commercial airliners.

ALLERGY AND ASTHMA

beWELL
http://beWELL.com/hic/allergy/
Medical and health information for the consumer. Specific pages on allergy.

E-SITE
http://www.latexallergyhelp.com/dental.htm
Latex allergies and exposures in the dental office.

Healthtouch
http://www.healthtouch.com/level1/leaflets/aan/aan103.htm
Immunotherapy (allergy shots) explained. Provided by Allergy and Asthma Network/Mothers of Asthmatics, Inc.

Sinus and Allergy Site

http://www.sinuses.com

Symptoms and treatment of sinusitis and other sinus diseases, plus interrelated problems of allergy and asthma. Also includes images of endoscopic image-guided surgery.

ARTICLES - SPECIFIC

American Journal of Medicine

http://www.ncbi.nlm.nih.gov/htbin-post/Entre

Series of articles on CFS in September 1998 supplemental issue.

Characteristic Anomalies in CFS Urine Profiles

www.newcastle.edu.au/department/bi/birjt/cpruis/urine.html

City Creates Havoc with Kalliss Family

http://www.telusplanet.net/public/dkalliss/index.html

Home page by the Kalliss family in Canada, recounting their struggle with the father's employer. Regardless of the truth of the facts, this scenario is all too common. Read and learn.

Interesting Canadian Links to:

- Occupational Health Services
- Alberta Labour Information
- Fraud and Bad Faith
- Environmental Illness Society of Canada
- Canadian Centre for Occupational Health & Safety
- How Workplace Chemicals Enter The Body

Dust Mite Allergy

http://www.medscape.com/Medscape/RespiratoryCare/1998/v02.n06/mrc3062.mcne-01.html

-also-

http://www.medscape.com/Medscape/public/MP/98/1120.html

Everyday Exposure to Toxic Pollutants

http://www.sciam.com/1998/0298issue/0298ott.html

From *Scientific American's* Web page.

hep-c-alert

www.hep-c-alert.org/oped0518.html

Hepatitis is much more accepted than CFS, MCS, EI, FM, etc. Yet patients feel they have difficulty with public health departments and health care professionals. Interesting to compare.

Indoor Air Quality Review (Monthly)

http://www.iaqpubs.com/ier.html

A U.S. District Court judge disallowed testimony from 12 physicians whose clients suffer from multiple chemical sensitivity (MCS).

http://www.iaqpubs.com/ier-rsrc/stories/story2.html

Story on EPA fact sheet on ozone generators.

http://www.iaqpubs.com/ier-rsrc/stories/story2.htm

Industrial hygienist discusses whom to believe: the occupants or the measurements?

Insult to Injury

http://www.pressdemo.com/workerscomp/

Workers' Compensation - Special Report

Lead-Based Paint (LBP) /TSCA Title IV

http://www.epa.gov/reg5foia/pb

A Joint Venture Between EPA and HUD

Lead-Based Paint (LBP) Disclosure Rule.

Not "Junk Science" but "Junk Journalism"

http://users.lanminds.com/~wilworks/ehnmcsrr.htm

John Stossel and ABC News on Multiple Chemical Sensitivity.

Original uploaded by permission of Albert Donnay of MCS Referral & Resources <www.mcsrr.org>

Perfume analysis

http://pw1.netcom.com/~bcb56/PerfAnalysis.htm

Lab results of two popular perfumes. See what's in them.

Plants and Sick Buildings

http://www.oxford.net/~steve/sick.htm

Based on NASA research.

Poor Treatment of the Chemically Exposed

http://www.mugc.cc.monash.edu.au/~degob1/goble/Acta/

Aust, Chemical Trauma Alliance (ACTA) is a nonprofit, volunteer organization concerned about people who have been harmed by chemicals. Includes information exchange, education, advice, counseling and referral. Promotes improving the status of its members through contact with governments.

Staudenmeyer, Herman - newspaper article about

http://bcn.boulder.co.us/media/colodaily/97/zmay16/CHEM16E.html

His book is included in that portion of the resources. He is controversial, at least in terms of the typical reader of this book. Also, it is important to know what the opposition has to say and to learn from their experiences. This and the following article contain a good flavor of the debate.

- also -

http://www.hcrc.org/contrib/acsh/booklets/mcsdoc.html

- also -

http://community-care.oaktree.co.uk/news/mcs0002.txt

A rebuttal to one of the articles.

Today's "Toxic News for the Net"

http://www.epa.gov/opptintr/oppt_nb.txt

Daily updates from the OPPT Library.

Toxicmold.com

http://www.toxicmold.com/

Home page and web site by a family in Phoenix whose personal, financial and physical life was altered by exposure to the toxic molds and mycotoxins of Stachybotrys atra, Penicillium and Aspergillus. They feel their situation was worsened by dealings with their insurance company. Again, regardless of the facts, this scenario is all too common.

Why Johnny Can't Breathe

http://users.lanminds.com/~wilworks/ehnmcsrr.htm

by Hank Hoffman of the New Haven Advocate.

CHRONIC FATIGUE, CFIDS, FIBROMYALGIA AND ME

AFSA

http://www.afsafund.org/

A nonprofit organization dedicated to research, education and patient advocacy for fibromyalgia syndrome (FMS) and chronic fatigue syndrome (CFS).

The CFIDS Association of America

http://www.cfids.org

Advocacy, Information, Research and Encouragement for the CFIDS Community. This is a 501(c)3 organization.

CFS-NEWS

http://www.cais.net/cfs-news

An electronic newsletter published by Roger Burns at no charge. See instructions for e-mail delivery. Numerous discussion groups and services. Since 1992. Subscription list of over 5,000 in 50 countries.

CFS Radio

http://members.aol.com/rgm1/private/transcr.htm

Previous transcripts can be viewed at:

http://www.cfsaudio.4biz.net/cfsradio.htm

Co-Cure

http://www.co-cure.org

Archives at http://listserv.nodak.edu/archives/co-cure.html

Its goal is furthering cooperative efforts of patients, care-givers and advocates towards finding the cure for chronic fatigue syndrome (CFS) and fibromyalgia FMS.

- also -

http://www.co-cure.org/schopflocher/AISH2B.htm

Article: The Chronic Fatigue Syndrome:
Qualification for Disability Benefits
Donald Schopflocher, Ph.D. (Psychology)
M.E/ C.F.S. Society of Edmonton

The Chronic Syndrome Support Association, Inc. (CSSA)

http://www.shore.net/~cssa

A nonprofit corporation. Its newsletter is *The Syndrome Sentinel.* Links to pages of personal info and political action.

Miningco
http://chronicfatigue.miningco.com/
Less obvious search engine for CFS.

National ME/FM Action Network of Canada
ww3.sympatico.ca/me-fm.action/

Parents and Young Persons with CFIDS
http://www.ypwcnet.org
For Young Persons with CFIDS (YPWCs).

Scientific References for CFS
http://www.networx.com.au/mall/cfs/data/
Over 1,100 papers categorized into 58 topics.

WECAN
http://www.cfids-me.org/wecan/

International on-line advocacy organization for people with chronic fatigue syndrome (CFS), chronic fatigue and immune dysfunction syndrome (CFIDS), or myalgic encephalomyelitis (ME).
WECAN newsletters can be found at:
http://www.cfids-me.org/wecan/inscom.html
The CFIDS/M.E. Information Page is at:
http://www.cfids-me.org/
More WECAN information at:
http://www.community-care.org.uk/wecan/
Insurance Committee's Web site:
http://www.cfids-me.org/wecan/inscom.html

CHRONIC ILLNESS - COPING

http://indigo.ie/~fubbs/
Ann O'Connell's Strategy for Coping — Irish Style — for chronic pain in her back and legs. How others cope with other conditions.

COSMETICS AND FRAGRANCE

Cosmetic Ingredients.

http://www.cosmeticmall.com/cm/html/CG_cos_ingred.html

Cosmetic Malls: Cosmetic Ingredient Dictionary.

Cosmetic Ingredients

http://www.ballbeauty.com/ingredie.htm

Ball Beauty Supply Homepage is a commercial site. This page for ingredients is worthwhile.

EHN — Fragrance Free — e-mail only

e-mail Wkfragfree@aol.com

-also - The Canary Club for chemically sensitive children and children who have chemically sensitive relatives.

Fragranced Products Information Network

http://www.ameliaww.com/fpin/fpin.htm

Grass-roots effort to educate about the chemicals used in, and the health effects of, fragranced products.

- also -

http://www.ameliaww.com/fpin/overview.htm

An overview of FPIN with links are provided to PubMed.

Olfactory Research Fund

www.olfactory.org

Find out about the industry side of the story. They claim that recent research studies show that fragrances can reduce stress, elevate mood and enhance sleep.

Perfume Analysis

http://pw1.netcom.com/~bcb56/PerfAnalysis.htm

Two popular perfumes were analyzed by a professional lab. See what was in them.

Safe, Unscented Products for MCS

http://www.sonic.net/daltons/melissa/unscent.html

Provided by Bastien, Sheila, Ph.D.

DENTAL

I have not mentioned dental issues in this book. However, it is a critical issue for some sufferers and you should at least be aware of the issues involved. Simply put, the use, storage and disposal of mercury is strictly controlled. Foods containing mercury are banned or tightly controlled at very low levels. Waste disposal sites cannot legally accept mercury — including dental amalgams that have been removed from a patient's mouth — unless it is specifically designed and certified to do so. Yet, there is one location where mercury is considered safe — inside the human mouth.

American Dental Association
http://www.ada.org/index.html

Dental Amalgam Mercury Syndrome Inc. (DAMS, Inc.)
http://www.lib.ci.tucson.az.us/orgfile/dental.htm
Education about possible effects of mercury dental fillings and other toxic conditions related to dentistry.

Dental X Change
http://dentalxchange.com/
A service and knowledge center for the enrichment of dentistry.

e-site
http://www.latexallergyhelp.com/dental.htm
Latex allergies and exposures information and links.

Foundation for Toxic Free Dentistry (FTFD)
<no web page - but you may want to know about this>
and a mercury-free dentist near you.

Call (800-331-2303), or send a
SASE to, PO Box 608010,
Orlando, FL 32860-8021.

Internet Dentistry Resources
http://indy.radiology.uiowa.edu/Beyond/Dentistry/sites.html
University of Iowa College of Dentistry - Links.

Mercury Free and Healthy

http://www.amalgam.org/Report: Prepared November 1998 by DAMS Inc. along with Consumers for Dental Choice. A Project of the National Institute for Science, Law and Public Policy, 1424 16th Street, NW, Suite 105, Washington, DC 20036

Minamata Disease

http://vest.gu.se/~bosse/Mercury/Mouth/amalgamnews.html

Amalgam-related news under the premise that the core problem is connected to how knowledge of various types is treated and the unreasonableness of having one kind of risk estimation for the environment, and another kind for what is put into people's mouths.

Lists of references with links to abstracts with headings of:

Man's mercury burden
Immunological reactions
Occupational hazards: dentistry
Amalgam removal
Amalgams: material aspects
Mercury pollution: dentistry

a paper: "The symbiosis between the dental and industrial communities and their scientific journals."

http://vest.gu.se/~bosse/Mercury/Ulf/symbiosis.html

a paper: "On the Instability of Amalgams"

http://vest.gu.se/~bosse/Mercury/Ulf/Instab/contents.html

ARTICLE: "Is Mercury Toxicity an Autoimmune Disorder?"

http://www.thorne.com/townsend/oct/mercury.html

DOCTOR SEARCH

DoctorNet

http://www.doctornet.com/

Interactive medical network for physicians and the public. Keyword search to find physicians. Can also browse by specialty and location.

MedSeek

http://medseek.com/

Database of more than 280,000 physicians throughout the U.S. whose home page is linked to this site. Both patients and doctors can search by specialty, geographic area or name.

Mental Health Net

www.cmhc.com

Disorders, articles, drugs with links. And a searchable Yellow Pages of more than 1,000 clinicians in the U.S.

E-MAIL RESOURCES

The Canary Club

e-mail Wkfragfree@aol.com

For chemically sensitive children and children who have
chemically sensitive relatives.

Fragrance Free

e-mail Wkfragfree@aol.com

MCS-CI (chemical injury)

http://www.ameliaww.com/MCS-CI/

The Web site is for info only and instruction on how to join the e-mail list.

MCS-IMMUNE-NEURO MCS

e-mail MCS-IMMUNE-NEURO@MAELSTROM.STJOHNS.EDU

Chemical injury support

Private Library of Articles

e-mail SWachsler@aol.com.

Articles for download and a list of articles available for free by e-mail.

ENVIRONMENTAL ISSUES

Blazing Tattles

http://www.concentric.net/~blazingt

"C. W. Gilbert" <blazing@igc.apc.org>

Linking chemical injury and health to exposure by pollution from weather and ecosystems and how that pollution affects the weather and ecosystem.

Environmental Defense Fund

http://www.scorecard.org

Information on toxic releases in communities in all 50 states accessed by zip code, county, city, state, chemical, or map.

The Green Disk

http://www.igc.org/greendisk

A journal of contemporary environmental issues published every two months on a computer disk in IBM and Mac formats.

Mission-Possible

http://www.dorway.com/possible.html

Martinni, Betty, e-mail Mission-Possible-USA@Altavista.net

http://www.holisticmed.com/aspartame/ Aspartame issues.

http://www.focusnewsletter.org/aspartam.htm Aspartame newsletter.

http://www.notmilk.com Bovine growth hormone issues.

The National Environmental Trust

http://www.envirotrust.com/

Public education campaigns about environmental issues facing Congress. Nonprofit. Addressing health, safety and environmental protections for food, air and drinking water safety, global climate change, public right-to-know policies and endangered species protection.

EPD - ENZYME POTENTIATED DESENSITIZATION

EPD is another specific topic not discussed in this book. However, many of my clients have significant improvement using this method. Find out more at these resources.

EPD HomePage
http://www.epdallergy.com/
W.A. Shrader, Jr., M.D., Principal Investigator, EPD IRB Study. A general summary of EPD immunotherapy.

EPD FAQ
http://www.dma.org/~rohrers/allergy/epd_faq.htm

EPD and Allergy Links
http://www.dma.org/~rohrers/allergy/allergy.htm

EPD Mailing List Archive
http://www.dma.org/~rohrers/allergy/epd_srch.htm

American EPD Society
www.epdallergy.com
Basic intro and references. For a list of EPD doctors in your local, call (505) 984-0004

FOOD SAFETY

Mission-Possible
http://www.notmilk.com
Exposing bovine growth hormone
http://www.holisticmed.com/aspartame/
Aspartame FAQ and info.

The ARK Institute
http://www.arkinstitute.com
Are people getting sick from the world's food supply?

Mothers for Natural Law

http://www.safe-food.org

Its original purpose was to transform societal problems into simple, practical actions for everyone. They have now made genetic engineering their only focus

Lists of brands and foods that have not been genetically engineered at

http://www.safe-food.org/-consumer/brands.html

GOVERNMENT ORGANIZATIONS AND SITES

Agency for Toxic Substances and Disease Registry - ATSDR

http://atsdr1.atsdr.cdc.gov:8080/alerts/961213.html

Centers for Disease Control and Prevention (CDC)

http://www.cdc.gov/

http://www.cdc.gov/ncidod/diseases/cfs/cfshome.htm

http://www.cdc.gov/nceh/meetings/1999/gulfwar/default.htm

For Gulf War info.

http://www.cdc.gov/epo/mmwr/mmwr.html

Morbidity and Mortality Weekly Report

http://www.cdc.gov/travel/travel.html.

Travel advisories

Consumer Product Safety Commission (CPSC)

http://www.cpsc.gov/talk.html

CPSC is an independent federal regulatory agency for consumer products safety. Go to "Talk to us" to report unsafe products.

EPA - Compliance Assistance Centers

Agriculture www.epa.gov/oeca/ag

Automotive Service & Repair www.ccar-greenling.org

Chemical Industry www.chemalliance.org

Local Governments www.lgean.org

Metal Finishing www.nmfrc.org

Paints & Coatings www.paintcenter.org

Printing www.pneac.org

Transportation www.transource.org

EPA - Office of Pollution Prevention and Toxics
http://www.epa.gov/opptintr/

Freedom of Information (FOI) with FDA:
http://www.fda.gov:80/opacom/backgrounders/foiahand.html
Instructions on how to file.

Government Information Xchange
http://www.hhs.gov/search/
Search ALL Federal Health and Human Services Agencies.

Grateful Med
http://igm.nlm.nih.gov.index.html
Free searches and links to other health literature.

Hazardous Materials Safety - HAZMAT
http://hazmat.dot.gov
Department of Transportation's Research and Special Programs Administration. National safety program for the transportation of hazardous materials by air, rail, highway and water.

Health and Human Services (DHHS)
http://www.dhhs.gov/

Medlars
http://www.nlm.nih.gov/databases/medlars/html
Search link for the pink-book scientific medical literature (listed in the back of the pink book).

MedLine
http://www.ncbi.nlm.nih.gov/PubMed/

MedlinePlus
http://medlineplus.nlm.nih.gov/medlineplus/
National health info for consumers.

MedWatch

http://www.fda.gov/medwatch/safety

FDA Medical Products Reporting Program. Safety information and postmarketing surveillance of medical products.

National Center for Environmental Health (NCEH)

http://www.cdc.gov/nceh/ncehhome.htm

National Library of Medicine

http://www.nlm.nih.gov/

With appropriate accounts, can access databases MEDLINE, AIDSLINE and TOXNET.

PubMed

http://www.ncbi.nlm.nih.gov/PubMed/

National Library of Medicine's search service. Nine million citations in MEDLINE and Pre-MEDLINE with links.

Safety & Health Internet Sites

http://www.osha.gov/safelinks.html

Links supplied by OSHA (Occupational Safety & Health Administration) <http://www.osha.gov>

I am including most of the list from their Web site. Some may not seem immediately relevant and others appear elsewhere in this Appendix. But I am including them because government agencies and departments are so massive and complex, and knowledge about current public health and safety is so critical to making good decisions. Plus, they have some unusual ones outside of the federal government itself.

American Academy of Physician Assistants in Occupational Medicine

http://www.aapa.org/paom.htm

American Association of Occupational Health Nurses

http://www.aaohn.org

American College of Occupational and Environmental Medicine (ACOEM)

http://www.acoem.org

American College of Preventive Medicine (ACPM)
http://www.acpm.org/

American Conference of Governmental Industrial Hygienists
http://www.acgih.org
American Public Health Association (APHA)
http://www.apha.org

American Red Cross
http://www.redcross.org

Agency for Toxic Substances and Disease Registry
http://ATSDR.atsdr.cdc.gov:8080/atsdrhome.html

American Industrial Hygiene Association
http://www.aiha.org/

American Society of Safety Engineers
http://www.ASSE.org/

Army Industrial Hygiene
http://chppm-www.apgea.army.mil/Armyih/

Canadian Centre for Occupational Health and Safety
http://www.ccohs.ca/

CCOHS Safety Related Internet Resources
http://www.ccohs.ca/Resources/hshome.htm

CDC (Centers For Disease Control And Prevention)
http://www.cdc.gov

Denison University, Campus Security & Safety
http://www.denison.edu/sec-safe/

Duke University Occupational & Environmental Medicine
http://occ-env-med.mc.duke.edu/oem

Eastern Washington University Center for Farm Health and Safety
http://www.farm.ewu.edu/

Ergo Web
http://tucker.mech.utah.edu

MSDS On-line from University of Utah
gopher://atlas.chem.utah.edu:70/11/MSDS

USDOL Mine Safety And Health Administration
http://www.msha.gov

MSU Radiation, Chemical & Biological Safety
http://www.orcbs.msu.edu

NIOSH (National Institute For Occupational Safety And Health)
http://www.cdc.gov/niosh/homepage.html

National Safety Council
http://www.nsc.org

Pan American Health Organization
http://www.paho.org/

Rocky Mountain Center for Occupational and Environmental Health
http://rocky.utah.edu

University of California, Irvine Health Promotion Center
http://www.socecol.uci.edu/~socecol/depart/research/hpc/hpc.html

University of Iowa Institute for Rural and Environmental Health
http://info.pmeh.uiowa.edu

University of Kansas School of Allied Health
http://www.kumc.edu/SAH/

University of London Ergonomics and Human Computer Interaction
http://www.ergohci.ucl.ac.uk/

US Department of Health and Human Services

http://www.os.dhhs.gov

Utah Safety Council

http://www.ps.ex.state.ut.us/sc/usc.htm

UVA's Video Display Ergonomics page

http://www.virginia.edu/~enhealth/ERGONOMICS/toc.html

World Health Organization

http://www.who.ch

Statistics

www.fedstats.gov

Statistics from more than 70 federal agencies.

Thomas Register

http://thomas.loc.gov/

Operated by the Library of Congress. Access to full text of the Congressional Record and bill database, House and Senate Roll Call votes. Daily updates of floor activity.

ToxFAQs

http://ATSDR.atsdr.cdc.gov:8080/toxfaq.html

Agency for Toxic Substances and Disease Registry — ATSDR

Alphabetical Index of ToxFAQs.

HOUSING AND MATERIALS

Anderson Labs

http://andersonlaboratories.com/alweb4.htm

Health effects of indoor air pollution with toxic chemicals. One of the few toxics labs doing paradigm breaking research, rather than limiting themselves to regulatory compliance.

Austin Green Builder Program

http://www.greenbuilder.com/sourcebook/FinishesAdhesives.html

Sustainable building sourcebook on paints, finishes, etc.

MCS Housing Resources

http://www.thegarden.net/mcs/

Forum for housing issues faced by people with multiple chemical sensitivities (MCS). Excellent links.

JOURNALS AND PUBLICATIONS

AAA Alternative Health Tips

www.iwr.com/ezine/issues.htm

A weekly ezine for alternative health methods of living.

American Journal of Medicine

http://www.ncbi.nlm.nih.gov/

Indoor Air Quality Updates - IAQU

http://www.cutter.com

JAMA - Journal of the American Medical Association

http://www.ama-assn.org/

Plus all other AMA publications. Access to the Journals.

Journal of Immunology Cutting Edge Papers

http://www.jimmunol.org

Rapid communications of novel research and current issues in immunology.

Journal of Neurology, Neurosurgery, and Psychiatry

http://www.pslgroup.com/dg/6bbee.htm

Article: "The degree of functional incapacity resulting from chronic fatigue syndrome parallels patients with HIV, multiple sclerosis, Alzheimer's disease and stroke."

The Lancet

http://www.thelancet.com/

National Environmental Trust

http://www.envirotrust.com/glreport01.html

Article - Hormone Disrupters in the Great Lakes Region.

New England Journal of Medicine on Asthma

http://www.nejm.org/collections/asthma/TOC/1.htm

Abstracts of original articles and special articles.

NY Times Syndicate Medical Tribune

http://www.medtrib.com/

NewsFile

http://www.newsfile.com

Medical Newsletters' Weekly Top News Stories from AIDS Weekly, Antiviral Weekly, Blood Weekly, Cancer Biotechnology Weekly, Emerging Pathogen Weekly, Gene Therapy Weekly, etc.

Priory Lodge

http://www.priory.com/

Network of interrelated biomedical journals, owned and administered by Priory Lodge Education Ltd. Active site, which claims over 600,000 hits per month.

Science News

http://www.sciencenews.org/sn_arc97/

Scientific American

http://www.sciam.com/

Washington Post

http://www.washingtonpost.com/wp-srv/

LEGAL

"Insult to Injury"

http://www.pressdemo.com/workerscomp/

Workmans' Compensation-Special Report

Indoor Environment Review

http://www.iaqpubs.com/ier.html

Testimony denied from physicians about people suffering from multiple chemical sensitivity (MCS).

MASSAGE

American Massage Therapy Association

http://www.amtamassage.org/

Frequently Asked Questions (FAQ), and other info on types of massage, benefits and how to find a massage therapist.

MEDICATIONS AND DRUGS

Healthtouch

http://www.healthtouch.com/level1/p_dri.htm

Information about more than 7,000 prescription and over-the-counter medications,

Chemical Searching (includes drugs)

http://chemfinder.camsoft.com/

Chemical Information Sites Index

http://chemfinder.camsoft.com/siteslist.html

Indigent Prescription Programs

http://www.li.net/~edhayes/indigent.html

Prescription medications free of charge to physicians whose patients might not otherwise have access to necessary medications.

also, many links at

http://www.li.net/~edhayes/ed.html

MEDICINE — ALTERNATIVE AND COMPLEMENTARY

Alternative Medicine HomePage

http://www.pitt.edu/~cbw/altm.html

Links to unconventional, unorthodox, unproven or alternative, complementary, innovative and integrative therapies.

The American Association of Naturopathic Physicians

http://healthy.net/pan/pa/Naturopathic/aanp

Disease and health info, plus curricula for naturopathic medical schools. Links.

Search site

http://WWW.MedWeb.Emory.Edu/MedWeb/

Sinus Survival

http://www.sinussurvival.com/

A book and program by Dr. Ivker. Includes a Q&A section.

Weil, Dr. Andrew

cgi.pathfinder.com/drweil/

The best-selling author and lecturer on alternative medicine answers your questions.

The Yoga Site

www.yogasite.com

MEDICINE AND HEALTH - CONTROVERSIAL

Environmental Sensitivities Research Institute - ESRI

http://www.esri.org

A 501(c)(3), nonprofit scientific and educational organization supporting sound scientific and medical research information, and the compilation and dissemination of information about environmental intolerance issues. *ESRI is primarily sponsored by its member organizations.*

The controversy intensifies when you find out who those member organizations are. It pays to know the background of any organization, including those in alternative fields.

The Federation of State Medical Boards

http://www.fsmb.org/HealthFraud.htm

This report on health fraud was accepted by the FSMB as policy in April 1997. You may want to know what it says.

Not "Junk Science" but "Junk Journalism"

http://users.lanminds.com/~wilworks/ehnmcsrr.htm

John Stossel and ABC News on multiple chemical sensitivity.

(Original on the Internet download by permission of AlbertDonnay MCS Referral & Resources).

There is a lot of talk about "junk science" and the harm caused by it. Even the U.S. Supreme Court has studied its effect in law. It is a very real concern, much as health fraud and quackery are.

But the other side of the coin is when and how the media gets the information they report. Several resources in this book talk about the "closed information systems" between the pharmaceutical companies and doctors, the dental supply manufacturers and the dental boards. Also included are the professional and regulatory agencies that monitor health practitioners.

We are all familiar with how the tobacco industry used "science" for decades to defend against claims of legal liability. And that was not much different from the asbestos industry and the silicon implant industry. Could it also be happening with the chemical industry, the electrical power industry, in dentistry and in medicine?

Albert Donnay has aggressively challenged the "pharmaceutical-chemical-regulatory-media complex" with this article and his MCS Referral & Resources organization (www.mcsrr.org). Likewise, the media aggressively defend their stand.

This is exciting reading about an issue all of us need to know more about and understand more clearly. It gets back to the issue at the beginning of this book: "Whom do *you* rely on for *your* decisions and how do *you* get the accurate information *you* need to make *your* decisions about *your* health and life?"

QuackWatch

http://www.quackwatch.com

Your Guide to Health Fraud, Quackery and Intelligent Decisions. Topics include:

How Should It Be Defined?

How It Harms

How It Sells

25 Ways to Spot It

"Health Freedom"

More Ploys That May Fool You

Common Misconceptions

Why People Are Vulnerable

Why It Persists
Victim Case Reports
Nonvictim Case Reports
Ten Ways to Avoid Being Quacked
Signs of a "Quacky" Web Site
Pro-Quackery Legislation
Incisive Comments
Propaganda Techniques Related to Environmental Scares
Why Strong Laws Are Needed to Protect Us
Some Notes on the Nature of Science
Why Bogus Therapies Often Seem to Work

While you may agree with the concept and the need to fight Quackery, the defense against it is almost always based on statistical distribution measurements of large groups. Is it possible to have due diligence without sacrificing the individual's actuality of having an individual experience?

Staudenmeyer — newspaper article on
http://bcn.boulder.co.us/media/colodaily/97/zmay16/CHEM16E.html
http://www.hcrc.org/contrib/acsh/booklets/mcsdoc.html
- also - This rebuttal letter at:
http://community-care.oaktree.co.uk/news/mcs0002.txt

MEDICINE - GENERAL

Don't expect to find much, if any, information about your *individual* health problems on any of these sites. This category is included to support one of the themes of this book — which is to know the public health arena as well as your individual health requirements.

However, these sites do present an opportunity to educate the health care professionals. Most actively solicit questions for difficult health problems. So, ask them some questions. Such as:

Where on your Web site can I find information about Chronic Fatigue Syndrome, Fibromyalgia or Multiple Chemical Sensitivity? Especially since the Center for Disease Control (CDC) has had a diagnostic for Chronic Fatigue Syndrome (CFS) for several years. And Multiple Chemical Sensitivity (MCS) is an accepted category by Social Security for disability. Furthermore, several other agencies accept it under the Americans with Disabilities Act. So where is the information?

What do I do when I feel sick at the office but the doctors insist nothing is wrong with me? And the building management insists that they are in compliance with all regulations, including OSHA?

How do you suggest I handle a dysfunctional medical relationship?

How do I explain to my child's teacher that her heavy perfume causes his rash or hyperactive behavior?

How do I test for allergens in my home or at my child's school?

How do I find a doctor who understands the role of exposure in allergies and asthma?

How do you recommend I find and use natural or herbal products instead of the expensive and often "toxic" pharmaceuticals typically prescribed?

These are just suggestions. Clarify your own needs and then formulate a question. If they don't know our needs, they will never realize that we have them. Ask the questions for the purpose of education. Who knows, one of them may be able to help you. And maybe they will be open to learning about a whole other world of health care.

Allergy, Asthma & Immunology Online

http://allergy.mcg.edu/

American Academy of Allergy, Asthma & Immunology

http://www.aaaai.org/

American Medical Association

http://www.ama-assn.org/

Ask Alice

http://www.columbia.edu/cu/healthwise/

Professional answers to your health questions.

Cool Medical Sites

http://www.hooked.net/users/wcd/cmsotw.html

Dr. Koop's Community

www.drkoop.com

Ambitious site by C. Everett Koop, former U.S. Surgeon General. It has nearly 50 searchable categories of medicine, plus links and Q&A.

Health Daily Article

http://yourhealthdaily.com/

The day's breaking health news, searchable database and discussion groups. Links to other sites.

HealthFinder

www.healthfinder.com

Developed by the Department of Health and Human Services with cooperation from other federal agencies. A clearinghouse for information.

Mayo Clinic

http://www.mayohealth.org/

Medicine Net

www.medicinenet.com

Definitions for hundreds of diseases and treatments, along with prescription and over-the-counter drugs. Its "Ask the Experts" area has thousands of searchable Questions and Answers.

Mental Health Net

www.cmhc.com

Deals in disorders, articles and drugs with links. Includes a searchable Yellow Pages of more than 1,000 clinicians in the U.S.

NOAH-New York Online Access to Health

http://www.noah.cuny.edu/asthma/asthma.html

Personal Health Tracker

http://www.onhealth.com/hnews/9510/htm/latex.htm

Sign up for e-mail info on your requested topics.

MSDS - MATERIAL SAFETY DATA SHEETS

MSDS Search

http://www.msdssearch.com/

Some 87,000 hits a month for access to 100,000 MSDSs directly from the manufacturer, 500,000 from public data links to software and service providers and government MSDS.

MSDS sites on the Internet

http://www.ilpi.com/msds/index.chtml#Internet

Over 60 free sites to search.

University of Oregon - MSDS

http://chemlabs.uoregon.edu/Safety/MSDS.html

Good source for basic info.

MSDS Online

http://www.msdsonline.com/

Free online search tools to help you search the Internet for MSDSs and company contact information. Free MSDS document manager (MSDS Manager 2.1).

MSDS On-line

gopher://atlas.chem.utah.edu:70/11/MSDS

University of Utah site.

Grip.com

http://www.grip.com/msds.htm

Very comprehensive site.

Toreki, Rob

http://www.ilpi.com/msds/index.chtml

Extremely comprehensive listing of online sites.

MULTIPLE CHEMICAL SENSITIVITY AND ENVIRONMENTAL ILLNESS

Aust, Chemical Trauma Alliance - ACTA

http://www.mugc.cc.monash.edu.au/~degob1/goble/Acta/

Article: "Poor Treatment of the Chemically Exposed."

Blazing Tattles

http://www.concentric.net/~blazingt

"C. W. Gilbert" <blazing@igc.apc.org>

Linking chemical injury and health to exposure by pollution from weather and ecosystems and how that pollution affects the weather and ecosystem.

The Chemical Injury Information Network (CIIN)

http://biz-comm.com/CIIN/

A nonprofit organization focusing primarily on education, credible research into Multiple Chemical Sensitivities (MCS), and the empowerment of the chemically injured.

The Environmental Health Network (EHN) [of Calif.

http://users.lanminds.com/~wilworks/ehnindex.htm

EHN is a 501(C)3 nonprofit agency for those disabled by chemical injury and/or sick building syndrome. Uses access advocacy, peer counseling, newsletters, books and meetings.

Environmental Illness Society of Canada

http://www.cyberus.ca/eisc/.

Hamline University

http://www.hamline.edu/lupus

Articles explaining various aspects of lupus, plus links.

HEAL (Human Ecology Action League)

http://members.aol.com/HEALNatnl/index.html

A national organization with chapters throughout the U.S.

Immune List

http://www.best.com/~immune

Probably the best and most active e-mail-based list.

News headlines page

http://headlines.isyndicate.com/rcscript/?user=m3btrauc

Journal of Neurology, Neurosurgery, and Psychiatry

http://www.pslgroup.com/dg/6bbee.htm

Article: "The degree of functional incapacity resulting from Chronic Fatigue Syndrome parallels findings in patients with HIV, multiple sclerosis, Alzheimer's disease and stroke."

MCS-CI mailing list

http://www.ameliaww.com/MCS-CI/

MCS Housing Resources

http://www.thegarden.net/mcs/

On-line discussion forum for housing issues faced by people with multiple chemical sensitivities (MCS). Excellent links.

MCS-IMMUNE-NEURO MCS/Chemical Injury Support

E-mail at MCS-IMMUNE-EURO@MAELSTROM.STJOHNS.EDU

MCS Referral & Resources

http://www.mcsrr.org/

Albert Donnay - Professional outreach, patient support and public advocacy devoted to the diagnosis, treatment, accommodation and prevention of multiple chemical sensitivity disorders.

Mission-Possible

ORGANIZATIONS - PRIVATE OR NONPROFIT

AAEM - American Academy of Environmental Medicine:

http://www.healthy.net/pan/pa/NaturalTherapies/aaem/index.html

Its purpose is to assist the patient in uncovering the cause-and-effect relationship between their environment and their ill-health, and to help them learn to avoid those inciting factors.

More than 400 physicians, searchable, plus links.

American Academy of Pediatrics

www.aap.org

Child health issues and resources on education, advocacy, and research.

Environmental Illness Society of Canada

http://www.cyberus.ca/eisc/.

Environmental Defense Fund

http://www.scorecard.org

Information on toxic releases in communities in all 50 states accessed by zip code, count, city, state, chemical or map.

PESTICIDES

Agency for Toxic Substances and Disease Registry - ATSDR
http://ATSDR.atsdr.cdc.gov:8080/alerts/961213.html
National Alert
Illegal Use of Methyl Parathion Insecticide

AirTech/Chem-Tox
http://www.chem-tox.com/pesticides/pesticides.htm
Pesticide Health Effects Research
Abstracts of studies done on the health effects.

Northwest Coalition for Alternatives to Pesticides
www.efn.org/~ncap

Pesticide Action Network
www.panna.org/panna

Pesticide Management Education Program (PMEP)
http://pmep.cce.cornell.edu/
For Pesticide Active Ingredient Profiles see:
http://pmep.cce.cornell.edu/profiles/index.html

SCHOOLS AND CHILDREN

American Federation of Teachers - AFT
www.aft.org
Information on school environments, fact sheets on IAQ, asbestos and mold. "Healthy Schools" newsletter with guidelines to parents and staff.

American Lung Association
www.lungusa.org
Info about air quality and asthma with children. Includes a program "Open Airways for Schools" for teaching children and communities about environmental factors in schools, and how to improve school conditions.

American Public Health Association

www.apha.org

Includes the "Children's Environmental Health Fact Sheet."

Children's Environmental Health Network - **(CEHN)**

www.cehn.org

Children's health, including national reports on policy and research issues, educational programs, publications. Links.

Doris Rapp

www.drrapp.com/index.html

Indoor Air Diagnostics, Inc.

www.iaqinfo.com

Specific information and facts including school environments and hazards.

National Education Association - **NEA**

www.nea.org/resource/safe.html

Web site on safe schools.

National Parent Network of Disabilities

www.npnd.org

"Environmental Health News" school programs and child health research about current environmental issues. Also, reports by ASTHO - The Association of State and Territorial Health Officials.

National PTA's Environmental Action and Awareness Program

www.pta.org/programs

Articles on lead, radon, IAQ on children's health issues, environmental hazards. Newsletter.

Natural Resources Defense Council - **(NRDC)**

www.nrdc.org

Identifies worst environmental threats to children's health and learning, and their special vulnerability. Steps towards education and prevention.

New York Healthy Schools Network

http://www.hsnet.org/

Find out about 504 plans and i.e.p. Also, "Healthy Schools Survey Checklist" that you can print and use.

Physicians for Social Responsibility

www.psr.org

Report: "Healthy Children – Toxic Environments"

Lots of information and resources on school environments, effects on children's health and vulnerability to environmental threats, results from exposures, roles of other agencies and organizations.

SEARCH ENGINES AND DATA BASES

Accufind

http://www.nln.com/

18 search engines.

Alison Hunter Memorial Foundation

http://www.networx.com.au/mall/cfs/data/

More than 1,100 papers categorized into 58 topics of scientific references for CFS.

Avicenna Medline Access

http://www.avicenna.com/

Free searching after a moderate registration process. Offers simple and advanced searching.

Bill Bebout

http://www.evansville.net/~wbbebout

Impressive Web page with links by an aware individual.

Britannica, Encyclopedia

http://www.eblast.com/search.html

CambridgeSoft

http://www.camsoft.com/

Chemical Information Sites Index
http://chemfinder.camsoft.com/siteslist.html

Chemical Searching (includes drugs)
http://chemfinder.camsoft.com/

Community of Science, Inc. - Medline
http://www.cos.com/
Free searches for librarian users. Utilizes advanced features such as natural language and fuzzy logic.

Cool Medical Sites
http://www.hooked.net/users/wcd/cmsotw.html

The Environmental Defense Fund
http://www.scorecard.org
Toxic releases in communities in all 50 states.

Grateful Med
http://igm.nlm.nih.gov.index.html
Free searches and lots of links to other health literature.

HealthGate Medline
http://www.healthgate.com/HealthGate/MEDLINE/search.shtml
Free searches sponsored by advertisements. An intuitive interface that offers options for advanced users. Full text delivery is relatively expensive. Moderate enhancements for librarian or advanced searchers.

Healthtouch
http://www.healthtouch.com/level1/p_dri.htm
Information about more than 7,000 prescription and over-the-counter medications.

Helix Medline Access
http://www.helix.com/
Developed by GlaxoWelcome. Free for professionals after completion of a registration form.

Infotrieve

http://www.infotrieve.com/

Library services company offering full-service document delivery, databases on the web and a variety of tools.

Matilda search engine

http://www.aaa.com.au/submit/

The Med Engine

http://www.fastsearch.com/med/index.html

Hundreds of links to medical ininformation

Medlars

http://www.nlm.nih.gov/databases/medlars/html

Search link for the pink-book scientific medical literature.

MedLine

http://www.ncbi.nlm.nih.gov/PubMed/

- and -

http://www.healthgate.com/HealthGate/MEDLINE/search.shtml

http://www.healthgate.com/HealthGate/MEDLINE/search-adv.shtml

MedNab - M*N

http://www.medsitenavigator.com

High-quality medical and scientific information.

Medscape Medline Search

http://www.medscape.com/textSearch.form.fcgi

Free searches sponsored by advertisers, but requires a one-time membership registration.

MedSeek

http://medseek.com/

Directory to more than 280,000 physicians searchable by specialty, geographic area or name.

Miningco

http://chronicfatigue.miningco.com/

An excellent, high-quality search engine.

National Library of Medicine

http://www.nlm.nih.gov/

Ovid On Call - Medline and Medical Databases

http://preview.ovid.com/libpreview/

Advanced-fee-based and institutional interface. Full text e-mail delivery. Complex registration. Some 300 journals by the end of 1997.

Paper Chase Medical Literature Searching

http://www.paperchase.com/

Mesh expansion, intelligent relevance ranking, and selection based on multiple descriptors. Fee transaction based. Full text delivery by an 800 number.

Search ALL Federal Health and Human Services Agencies or the Government Information Xchange

http://www.hhs.gov/search/

Search Site for Alternative Medicine

http://WWW.MedWeb.Emory.Edu/MedWeb/

WORKMAN'S COMP & DISABILITY

Ragged Edge Magazine, March/April 1998

http://www.ragged-edge-mag.com

Covers the disability experience in America. Medical rationing, genetic discrimination, assisted suicide, long-term care, attendant services.

Workers' Compensation

http://www.pressdemo.com/workerscomp/

Special Report — "Insult to Injury."

Order Form

GMC Media

1811 South Quebec Way, #99
Denver, CO 80231-2671
or Toll Free
(877) 782-7878

Date:___________________________

Name: _________________________

Address: _______________________

____

City: __________________________

State, Zip ______________________

Phone (_____) _________________

1 through 3 additional copies of *Starting Points for a Healthy Habitat* are available for $34.95 each, plus $3.00 Shipping and Handling for the first copy. Shipping and Handling for additional copies in the same order is $1.00 each.

Quantity discounts are available to anyone wishing to resell the book.

A special consultant and seminar program is also available for medical doctors and other health care professionals.

Qty.	Description	Total
___	***Starting Points for a Healthy Habitat*** at **$34.95 each**	________
	3.8% Sales Tax for Colorado residents	________
	Postage and Handling	________
	TOTAL DUE	________

Payment: ❑ Check ❑ Money Order ❑ Visa ❑ MasterCard

Account #:____ Exp.Date_____

Signature ____ *(credit card requirement)*

Orders may be placed via: **Phone at toll free (877) 782-7878**
Faxed to (303)751-0416
or Mailed to 1811 So. Quebec Way, Suite 99
Denver, CO 80231-2671